Handbook of Medicinal Plants

Handbook of Medicinal Plants

Editor: Daniel Livingston

Callisto Reference

www.callistoreference.com

Callisto Reference,
118-35 Queens Blvd., Suite 400,
Forest Hills, NY 11375, USA

Visit us on the World Wide Web at:
www.callistoreference.com

ISBN: 978-1-64116-001-8 (Hardback)

Printed in China.

Cataloging-in-Publication Data

Handbook of medicinal plants / edited by Daniel Livingston.
p. cm.
Includes bibliographical references and index.
ISBN 978-1-64116-001-8
1. Medicinal plants. 2. Botany, Medical. 3. Materia medica, Vegetable. 4. Pharmacognosy.
I. Livingston, Daniel.
SB293 .H36 2018
633.88--dc23

Table of Contents

Preface

Medicinal herbs or medicinal plants have been used therapeutically for centuries. They have the chemicals which have pharmaceutical values and can been used to treat many ailments like skin diseases, cold, cough, body aches, fever, etc. The most common materials that are derived from medicinal plants are glycosides, terpenes, alkaloids, polyphenols, etc. This book is compiled in such a manner, that it will provide in-depth knowledge about the usage of medicinal plants. It elucidates new techniques and their applications in a multidisciplinary approach. For someone with an interest and eye for detail, this textbook covers the most significant topics in the field of medicinal plants. It will serve as a reference to a broad spectrum of readers.

Given below is the chapter wise description of the book:

Chapter 1- Medicinal plants are plants used in herbal medicine. These plants can have positive as well as harmful effects. Medicinal plant conservation, bioprospecting and traditional medicine are some of the topics related to medicinal plants that have been discussed in the following chapter.

Chapter 2- Hebe stricta is used to control diarrhea and dysentery whereas chamomile can cure inflammation and menstrual disorders. Laurelia novae-zelandiae, leptospermum scoparium, macropiper excelsum, celery and ballota nigra. The chapter offers an insightful focus, keeping in mind the complex subject matter.

Chapter 3- Medicinal plants have been used to treat patients in various traditions like traditional African medicine, traditional Chinese medicine and Ayurveda. Traditional Chinese medicine includes herbs such as Angelica sinensis, goji, Astragalus, Lingzhi mushroom, etc. The topics discussed in the chapter are of great importance to broaden the existing knowledge on medicinal plants.

Chapter 4- Humans have been using plants for medicinal practices for thousands of years. Herbal medicine is considered to be a form of alternative medicine in today's time. Decoction and zoopharmacognosy are some of the techniques used to make herbal medicine. This chapter will provide an integrated understanding of herbalism.

At the end, I would like to thank all those who dedicated their time and efforts for the successful completion of this book. I also wish to convey my gratitude towards my friends and family who supported me at every step.

Editor

1

An Introduction to Medicinal Plants

Medicinal plants are plants used in herbal medicine. These plants can have positive as well as harmful effects. Medicinal plant conservation, bioprospecting and traditional medicine are some of the topics related to medicinal plants that have been discussed in the following chapter.

Medicinal Plants

The bark of willow trees contains salicylic acid, the active metabolite of aspirin, and has been used for millennia to relieve pain and reduce fever.

Medicinal plants, medicinal herbs, or simply herbs have been identified and used from prehistoric times. Plants make many chemical compounds for biological functions, including defence against insects, fungi and herbivorous mammals. Over 12,000 active compounds are known to science. These chemicals work on the human body in exactly the same way as pharmaceutical drugs, so herbal medicines can be beneficial and have harmful side effects just like conventional drugs. However, since a single plant may contain many substances, the effects of taking a plant as medicine can be complex.

The earliest historical records of herbs are found from the Sumerian civilisation, where hundreds of medicinal plants including opium are listed on clay tablets. The Ebers Papyrus from ancient Egypt describes over 850 plant medicines, while Dioscorides documented over 1000 recipes for medicines using over 600 medicinal plants in *De materia medica*, forming the basis of pharmacopoeias for some 1500 years. Drug research makes use of ethnobotany to search for pharmacologically active substances in nature, and has in this way discovered hundreds of useful compounds. These include the common drugs aspirin, digoxin, quinine, and opium. The compounds found in plants are of many kinds, but most are in four major biochemical classes, the alkaloids, glycosides, polyphenols, and terpenes.

Medicinal plants

Medicinal plants are widely used to treat disease in non-industrialized societies, not least because they are far cheaper than modern medicines. The annual global export value of pharmaceutical plants in 2012 was over US$2.2 billion.

History

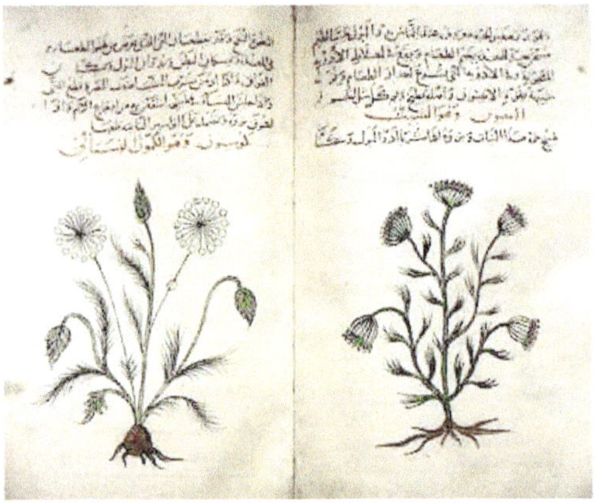

Dioscorides's 1st century *De materia medica*, seen here in a c. 1334 copy in Arabic, describes some 1000 drug recipes based on over 600 plants.

Prehistoric Times

Plants, including many now used as culinary herbs and spices, have been used as medicines from prehistoric times. Spices have been used partly to counter food spoilage bacteria, especially in hot climates, and especially in meat dishes which spoil more readily. Angiosperms (flowering plants) were the original source of most plant medicines. Human settlements are often surrounded by weeds useful as medicines, such as nettle, dandelion and chickweed. Humans were not alone in using herbs as medicines: some animals such as non-human primates, monarch butterflies and sheep ingest medicinal plants to treat illness. Plant samples from prehistoric burial sites are among the lines of evidence that Paleolithic peoples had knowledge of herbal medicine. For instance, a 60 000-year-old Neanderthal burial site, "Shanidar IV", in northern Iraq has yielded large amounts of pollen from 8 plant species, 7 of which are used now as herbal remedies, A mushroom was found in the personal effects of *Ötzi the Iceman*, whose body was frozen in the Ötztal Alps for more than 5,000 years. The mushroom was probably used to treat whipworm.

Ancient Times

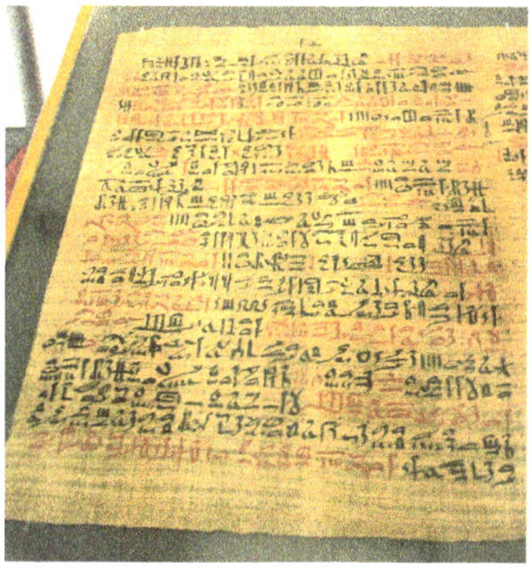

The Ebers Papyrus (*c*. 1550 BC) from Ancient Egypt describes the use of hundreds of plant medicines.

In ancient Sumeria, hundreds of medicinal plants including myrrh and opium are listed on clay tablets. The ancient Egyptian Ebers Papyrus lists over 800 plant medicines such as aloe, cannabis, castor bean, garlic, juniper, and mandrake. From ancient times to the present, Ayurvedic medicine as documented in the Atharva Veda, the Rig Veda and the Sushruta Samhita has used hundreds of pharmacologically active herbs and spices such as turmeric, which contains curcumin. The Chinese pharmacopoeia, the *Shennong Ben Cao Jing* records plant medicines such as chaulmoogra for leprosy, ephedra, and hemp. This was expanded in the Tang Dynasty *Yaoxing Lun*. In the fourth century BC, Aristotle's pupil Theophrastus wrote the first systematic botany text, *Historia plantarum*. In the first century AD, the Greek physician Pedanius Dioscorides documented over 1000 recipes for medicines using over 600 medicinal plants in *De materia medica*; it remained the authoritative reference on herbalism for over 1500 years, into the seventeenth century.

Middle Ages

In the Early Middle Ages, Benedictine monasteries preserved medical knowledge in Europe, translating and copying classical texts and maintaining herb gardens. Hildegard of Bingen wrote *Causae et Curae* ("Causes and Cures") on medicine. In the Islamic Golden Age, scholars translated many classical Greek texts including Dioscorides into Arabic, adding their own commentaries. Herbalism flourished in the Islamic world, particularly in Baghdad and in Al-Andalus. Among many works on medicinal plants, Abulcasis (936–1013) of Cordoba wrote *The Book of Simples*, and Ibn al-Baitar (1197–1248) recorded hundreds of medicinal herbs such as *Aconitum*, nux vomica, and tamarind in his *Corpus of Simples*. Avicenna included many plants in his 1025 *The Canon of Medicine*. Abu-Rayhan Biruni, Ibn Zuhr, Peter of Spain, and John of St Amand wrote further pharmacopoeias.

Early Modern

The Early Modern period saw the flourishing of illustrated herbals across Europe, starting with the 1526 *Grete Herball*. John Gerard wrote his famous *The Herball or General History of Plants* in 1597, based on Rembert Dodoens, and Nicholas Culpeper published his *The English Physician Enlarged*. Many new plant medicines arrived in Europe as products of Early Modern exploration and the resulting Columbian Exchange, in which livestock, crops and technologies were transferred between the Old World and the Americas in the 15th and 16th centuries. Medicinal herbs arriving in the Americas included garlic, ginger, and turmeric; coffee, tobacco and coca travelled in the other direction. In Mexico, the sixteenth century *Badianus Manuscript* described medicinal plants available in Central America.

Illustrated books of plants: *The Grete Herball*, 1526.

Phytochemistry

Many plants produce chemical compounds for defence against herbivores. These are often useful as drugs. The major classes of pharmacologically active phytochemicals are described below.

Alkaloids

Alkaloids are bitter-tasting chemicals, very widespread in nature, and often toxic. There are several classes with different modes of action as drugs, both recreational and pharmaceutical. Medicines of different classes include atropine, scopolamine, and hyoscyamine (all from nightshade), berberine (from plants such as *Berberis* and *Mahonia*), caffeine (*Coffea*), cocaine (*Coca*), ephedrine (*Ephedra*), morphine (opium poppy), nicotine (tobacco), psilocin (many psilocybin mushrooms), reserpine (*Rauwolfia serpentina*), quinidine and quinine (*Cinchona*), vincamine (*Vinca minor*), and vincristine (*Catharanthus roseus*).

The opium poppy *Papaver somniferum* is the source of the alkaloids morphine and codeine.

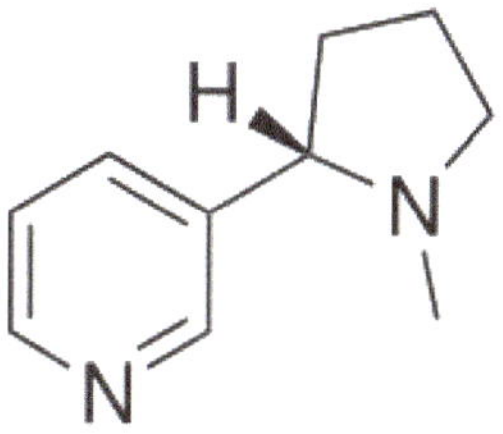

The alkaloid nicotine from tobacco binds directly to the body's Nicotinic acetylcholine receptors, accounting for its pharmacological effects.

Deadly nightshade, *Atropa belladonna*, yields tropane alkaloids including atropine, scopolamine and hyoscyamine.

Glycosides

Anthraquinone glycosides are found in the laxatives senna, rhubarb and *Aloe*.

The cardiac glycosides are powerful drugs from plants including foxglove and lily of the valley. They include digoxin and digitoxin which support the beating of the heart, and act as diuretics.

Senna alexandrina, containing Anthraquinone glycosides, has been used as a laxative for millennia.

The foxglove, *Digitalis purpurea*, contains digoxin, a cardiac glycoside.
The plant was used to treat heart conditions long before the glycoside was identified.

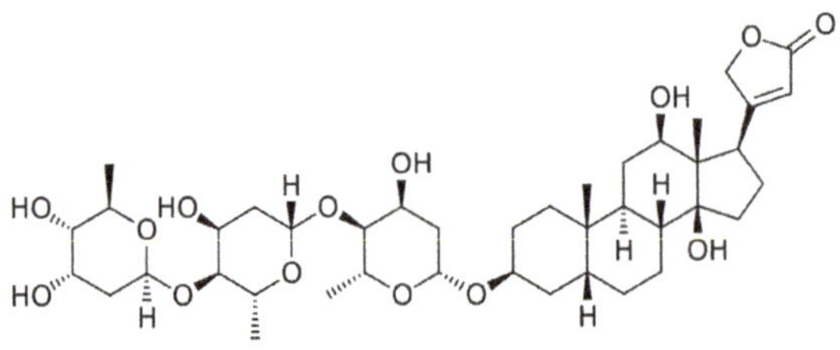

Digoxin is used to treat atrial fibrillation, atrial flutter and sometimes heart failure.

Polyphenols

Polyphenols of several classes are widespread in plants. They include the colourful anthocyanins, hormone-mimicking phytoestrogens, and astringent tannins. In Ayurveda, the astringent rind of the pomegranate is used as a medicine, while polyphenol extracts from plant materials such as grape seeds are sold for their potential health benefits despite the lack of evidence.

Plants containing phytoestrogens have been used for centuries to treat gynaecological disorders such as fertility, menstrual, and menopausal problems. Among these plants are *Pueraria mirifica*, kudzu, angelica, fennel, and anise.

Angelica, containing phytoestrogens, has long been used to treat gynaecological disorders.

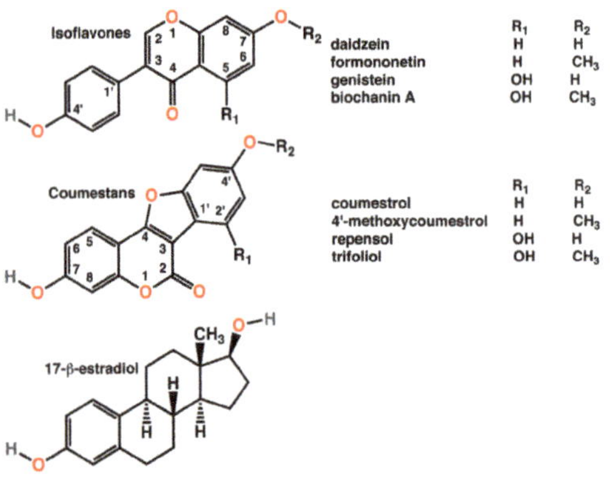

Polyphenols include phytoestrogens (top and middle), effective mimics of animal estrogen (bottom).

Terpenes

Terpenes and terpenoids of many kinds are found in resinous plants such as the conifers. They are strongly aromatic and serve to repel herbivores. Their scent makes them useful in essential oils, whether for perfumes such as rose and lavender, or for aromatherapy.

The essential oil of common thyme (*Thymus vulgaris*), contains the monoterpene thymol, an antiseptic and antifungal.

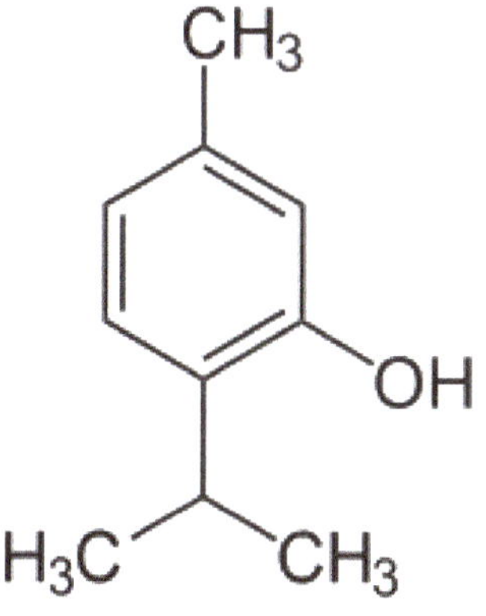

Thymol is one of many terpenes found in plants.

In Practice

Plant medicines are in wide use around the world. A herbalist's shop in the souk of Marrakesh, Morocco

Cultivation

Medicinal plants demand intensive management. Different species each require their own distinct conditions of cultivation. The World Health Organization recommends the use of rotation to minimise problems with pests and plant diseases. Cultivation may be traditional or may make use of conservation agriculture practices to maintain organic matter in the soil and to conserve water, for example with no-till farming systems. In many medicinal and aromatic plants, plant characteristics vary widely with soil type and cropping strategy, so care is required to obtain satisfactory yields.

Usage

Plant medicines are ubiquitous in pre-industrial societies, while some 7,000 conventional medicines such as aspirin, digitalis, opium, and quinine derive directly from traditional plant medicines, accounting for around a quarter of the modern pharmacopoeia. They are in general far cheaper, and many can be home-grown or picked for free. Further, pharmaceutical companies have made use of the herbal knowledge of indigenous peoples around the world to search for new drug candidates. In India, where Ayurveda has been practised for centuries, herbal remedies are the responsibility of a government department, AYUSH, under the Ministry of Health & Family Welfare. Traditional Chinese medicine makes use of a wide variety of plants, among other materials and techniques. Complementary and alternative medicines including herbal therapy are widely used in the Western world, for example by over a third of Americans.

Plant medicines including opiates, cocaine and cannabis have both medical and recreational uses. Different countries have at various times made some uses of drugs illegal, partly on the basis of the risks involved in taking psychoactive drugs.

Effectiveness

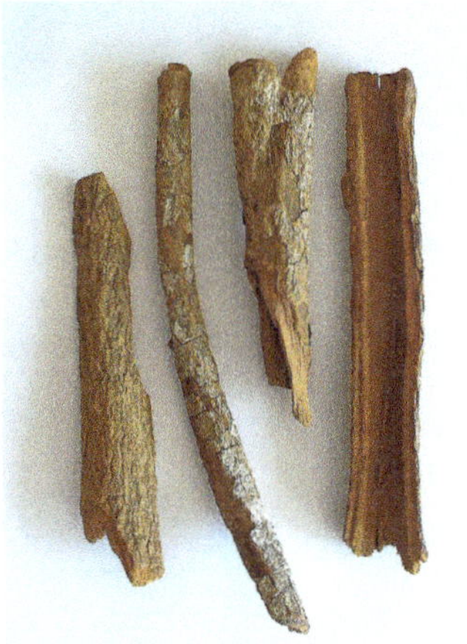

The bark of the cinchona tree contains the alkaloid quinine, a traditional treatment for malaria.

Plant medicines have often not been tested systematically, but have come into use informally over the centuries. By 2007, clinical trials had demonstrated potentially useful activity in nearly 16% of herbal medicines; there was limited in vitro or in vivo evidence for roughly half the medicines; there was only phytochemical evidence for around 20%; 0.5% were allergenic or toxic; and some 12% had effectively never been studied scientifically. According to Cancer Research UK, “there is currently no strong evidence from studies in people that herbal remedies can treat, prevent or cure cancer”. Modern knowledge of medicinal plants is being systematised in the Medicinal Plant Transcriptomics Database, which provides a sequence reference for the transcriptome of some thirty species. Hundreds of compounds have been identified using ethnobotany, investigating plants used by indigenous peoples. There is renewed interest in the discovery of therapeutically useful substances from medicinal plants.

Safety

Datura stramonium has been used to treat asthma, because it contains the alkaloid atropine, but it is also a powerful and potentially fatal hallucinogen.

Plant medicines can cause adverse effects and even death, whether by side-effects of their active substances, by adulteration or contamination, by overdose, or by inappropriate prescription. Many such effects are known, while others remain to be explored scientifically. There is no reason to presume that because a product comes from nature it must be safe: the existence of powerful natural poisons like atropine and nicotine shows this to be untrue. Further, the high standards applied to conventional medicines do not always apply to plant medicines, and dose can vary widely depending on the growth conditions of plants: older plants may be much more toxic than young ones, for instance. Pharmacologically active plant extracts can interact with conventional drugs, both because they may provide an increased dose of similar compounds, and because some phytochemicals interfere with the body's systems that metabolise drugs in the liver including the cytochrome P450 system, making the drugs last longer in the body and have a more powerful cumulative effect. Plant medicines can be dangerous during pregnancy. Since plants may contain many different substances, plant extracts may have complex effects on the human body.

Quality

Herbal medicines and supplements are not always tested for purity, and there is some concern about adulteration and inclusion of allergens such as soy and wheat in some products.

Traditional Medicine

Traditional medicine (also known as indigenous or folk medicine) comprises medical aspects of traditional knowledge that developed over generations within various societies before the era of modern medicine. The World Health Organization (WHO) defines traditional medicine as "the sum total of the knowledge, skills, and practices based on the theories, beliefs, and experiences indigenous to different cultures, whether explicable or not, used in the maintenance of health as well as in the prevention, diagnosis, improvement or treatment of physical and mental illness".

Traditional medicine in a market in Antananarivo, Madagascar

In some Asian and African countries, up to 80% of the population relies on traditional medicine for their primary health care needs. When adopted outside of its traditional culture, traditional medicine is often called alternative medicine. Practices known as traditional medicines include Ayurveda, Siddha medicine, Unani, ancient Iranian medicine, Iranian (Persian), Islamic medicine, traditional Chinese medicine, traditional Korean medicine, acupuncture, Muti, Ifá, and traditional African medicine. Core disciplines which study traditional medicine include herbalism, ethno-medicine, ethnobotany, and medical anthropology.

The WHO notes, however, that "inappropriate use of traditional medicines or practices can have negative or dangerous effects" and that "further research is needed to ascertain the efficacy and safety" of several of the practices and medicinal plants used by traditional medicine systems. The line between alternative medicine and quackery is a contentious subject.

Botánicas such as this one in Jamaica Plain, Massachusetts cater to the Latino community and sell folk medicine alongside statues of saints, candles decorated with prayers, lucky bamboo, and other items.

Traditional medicine may include formalized aspects of folk medicine, that is to say longstanding remedies passed on and practised by lay people. Folk medicine consists of the healing practices and ideas of body physiology and health preservation known to some in a culture, transmitted informally as general knowledge, and practiced or applied by anyone in the culture having prior experience. Folk medicine may also be referred to as traditional medicine, alternative medicine, indigenous medicine, or natural medicine. These terms are often considered interchangeable, even though some authors may prefer one or the other because of certain overtones they may wish to highlight. In fact, out of these terms perhaps only *indigenous medicine* and *traditional medicine* have the same meaning as *folk medicine*, while the others should be understood rather in a modern or modernized context.

Similarly, a home remedy is a treatment to cure a disease or ailment that employs certain spices, vegetables, or other common items. Home remedies may or may not have medicinal properties that treat or cure the disease or ailment in question, as they are typically passed along by lay people (a practice that has been facilitated in recent years by the Internet).

Usage and History

Classical History

In the written record, the study of herbs dates back 5,000 years to the ancient Sumerians, who described well-established medicinal uses for plants. In Ancient Egyptian medicine, the Ebers papyrus from c. 1552 BC records a list of folk remedies and magical medical practices. The Old Testament also mentions herb use and cultivation in regards to Kashrut.

Many herbs and minerals used in Ayurveda were described by ancient Indian herbalists such as Charaka and Sushruta during the 1st millennium BC. The first Chinese herbal book was the *Shennong Bencao Jing*, compiled during the Han Dynasty but dating back to a much earlier date, which was later augmented as the *Yaoxing Lun* (*Treatise on the Nature of Medicinal Herbs*) during the Tang Dynasty. Early recognised Greek compilers of existing and current herbal knowledge include Pythagoras and his followers, Hippocrates, Aristotle, Theophrastus, Dioscorides and Galen.

Roman sources included Pliny the Elder's *Natural History* and Celsus's *De Medicina*. Pedanius Dioscorides drew on and corrected earlier authors for his *De Materia Medica*, adding much new material; the work was translated into several languages, and Turkish, Arabic and Hebrew names were added to it over the centuries. Latin manuscripts of *De Materia Medica* were combined with a Latin herbal by Apuleius Platonicus (*Herbarium Apuleii Platonici*) and were incorporated into the Anglo-Saxon codex *Cotton Vitellius C.III*. These early Greek and Roman compilations became the backbone of European medical theory and were translated by the Persian Avicenna (Ibn Sīnā, 980–1037), the Persian Rhazes (Rāzi, 865–925) and the Jewish Maimonides.

Medieval and Later

Arabic indigenous medicine developed from the conflict between the magic-based medicine of the Bedouins and the Arabic translations of the Hellenic and Ayurvedic medical traditions. Spanish indigenous medicine was influenced by the Arabs from 711 to 1492. Islamic physicians and Muslim botanists such as al-Dinawari and Ibn al-Baitar significantly expanded on the earlier knowledge of materia medica. The most famous Persian medical treatise was Avicenna's *The Canon of Medicine*, which was an early pharmacopoeia and introduced clinical trials. The *Canon* was translated into Latin in the 12th century and remained a medical authority in Europe until the 17th century. The Unani system of traditional medicine is also based on the *Canon*.

Translations of the early Roman-Greek compilations were made into German by Hieronymus Bock whose herbal,published in 1546, was called *Kreuter Buch*. The book was translated into Dutch as

Pemptades by Rembert Dodoens (1517–1585), and from Dutch into English by Carolus Clusius, (1526–1609), published by Henry Lyte in 1578 as *A Nievve Herball.* This became John Gerard's (1545–1612) *Herball or General Hiftorie of Plantes*. Each new work was a compilation of existing texts with new additions.

Women's folk knowledge existed in undocumented parallel with these texts. Forty-four drugs, diluents, flavouring agents and emollients mentioned by Dioscorides are still listed in the official pharmacopoeias of Europe. The Puritans took Gerard's work to the United States where it influenced American Indigenous medicine.

Francisco Hernández, physician to Philip II of Spain spent the years 1571–1577 gathering information in Mexico and then wrote *Rerum Medicarum Novae Hispaniae Thesaurus*, many versions of which have been published including one by Francisco Ximénez. Both Hernandez and Ximenez fitted Aztec ethnomedicinal information into the European concepts of disease such as "warm", "cold", and "moist", but it is not clear that the Aztecs used these categories. Juan de Esteyneffer's *Florilegio medicinal de todas las enfermedas* compiled European texts and added 35 Mexican plants.

Martín de la Cruz wrote an herbal in Nahuatl which was translated into Latin by Juan Badiano as *Libellus de Medicinalibus Indorum Herbis* or *Codex Barberini, Latin 241* and given to King Carlos V of Spain in 1552. It was apparently written in haste and influenced by the European occupation of the previous 30 years. Fray Bernardino de Sahagún's used ethnographic methods to compile his codices that then became the *Historia General de las Cosas de Nueva España*, published in 1793. Castore Durante published his *Herbario Nuovo* in 1585 describing medicinal plants from Europe and the East and West Indies. It was translated into German in 1609 and Italian editions were published for the next century.

Colonial America

In 17th and 18th-century America, traditional folk healers, frequently women, used herbal remedies, cupping and leeching. Enslaved "root women" brought herbal knowledge and techniques of vaccination from Africa, infecting patients with a small bit of smallpox so they would build an immunity to the disease. In Appalachia, "granny women" of Irish and Scottish ancestry, acted as midwives and used herbs, laying on of hands, Bible verses, and other techniques to cure illnesses. Native American traditional herbal medicine introduced cures for malaria, dysentery, scurvy, non-venereal syphilis, and goiter problems. Many of these herbal and folk remedies continued on through the 19th and into the 20th century, with some plant medicines forming the basis for modern pharmacology.

Modern Usage

At the turn of the 20th century, folk medicine was viewed as a practice used by poverty-stricken communities and quacks. However, synthetic or biomedical products have been questioned by some parts of Western society, allowing for interest in natural medicines. The prevalence of folk medicine in certain areas of the world varies according to cultural norms. Some modern medicine is based on plant phytochemicals that had been used in folk medicine. Researchers state that many of the alternative treatments are "statistically indistinguishable from placebo treatments".

Knowledge Transmission and Creation

Indigenous medicine is generally transmitted orally through a community, family and individuals until "collected". Within a given culture, elements of indigenous medicine knowledge may be diffusely known by many, or may be gathered and applied by those in a specific role of healer such as a shaman or midwife. Three factors legitimize the role of the healer – their own beliefs, the success of their actions and the beliefs of the community. When the claims of indigenous medicine become rejected by a culture, generally three types of adherents still use it – those born and socialized in it who become permanent believers, temporary believers who turn to it in crisis times, and those who only believe in specific aspects, not in all of it.

Elements in a specific culture are not necessarily integrated into a coherent system, and may be contradictory. In the Caribbean, indigenous remedies fall into several classes: certain well-known European medicinal herbs introduced by the early Spaniard colonists that are still commonly cultivated; indigenous wild and cultivated plants, the uses of which have been adopted from the Amerindians; and ornamental or other plants of relatively recent introduction for which curative uses have been invented without any historical basis.

Rights of ownership may be claimed in indigenous medical knowledge. Use of such knowledge without Prior Informed Consent of or compensation to those claiming such ownership may be termed 'biopiracy'.

Folk Medicine

Curandera performing a *limpieza* in Cuenca, Ecuador

All cultures and societies have knowledge best described as folk medicine. Although there is large overlap, the denotative and connotative definitions differ. Folk medicine often coexists with formalized, education-based, and institutionalized systems of healing such as Western Medicine or Great traditional medicine systems like Ayurvedic, Unani medicine, and Chinese medicine, but is distinguishable from formalized or institutionalized healing systems.

Some examples of strong informal and to some degree institutionalized folk medicine traditions are: Traditional Chinese medicine, traditional Korean medicine, Arabic indigenous medicine (source of Unani medicine, along with ancient Greek medicine), Haitian folk medicine, Uyghur traditional medicine, Various African herbal folk remedies, Celtic traditional medicine (in part practiced by the Irish medical families), Japanese Kampō medicine, traditional Aboriginal bush medicine, Georgian folk medicine, and others.

Home Remedies

A home remedy (sometimes also referred to as a granny cure) is a treatment to cure a disease or ailment that employs certain spices, vegetables, or other common items. Home remedies may or may not have medicinal properties that treat or cure the disease or ailment in question, as they are typically passed along by laypersons (which has been facilitated in recent years by the Internet). Many are merely used as a result of tradition or habit or because they are effective in inducing the placebo effect.

One of the more popular examples of a home remedy is the use of chicken soup to treat respiratory infections such as a cold or mild flu, and according to one *in vitro* study, there may be benefit from this use. Other examples of medically successful home remedies include willow bark tea to cure headaches and fevers (willow bark contains salicylic acid, which is chemically similar to acetylsalicylic acid, also known as aspirin); duct tape to help with setting broken bones; and duct tape or superglue to treat plantar warts; and Kogel mogel to treat sore throat.

In earlier times, mothers were entrusted with all but serious remedies. Historic cookbooks are frequently full of remedies for dyspepsia, fevers, and female complaints.

Many European liqueurs or digestifs were originally sold as medicinal remedies. In Chinese folk medicine, medicinal congees (long-cooked rice soups with herbs), foods, and soups are part of the healing repertoire.

Many people also use aloe leaves to cure ailments of the skin.

Criticism

Safety Concerns

Although 130 countries have regulations on folk medicines, there are risks associated with the use of them. It is often assumed that because supposed medicines are herbal or natural that they are safe, but numerous precautions are associated with using herbal remedies.

Use of Endangered Species

Sometimes traditional medicines include parts of endangered species, such as the slow loris in Southeast Asia.

Endangered animals, such as the Slow loris, are sometimes killed to make traditional medicines.

Shark fins have also been used in traditional medicine, and although its use has not been proven, it is hurting shark populations and their ecosystem.

Medicinal Plant Conservation

Medicinal plant conservation strategies need to be understood and planned for based on an understanding of indigenous knowledge and practices. Many drugs contain herbal ingredients, and it has been said that 70–80% of the world's population relies on some form of non-conventional medicine and around 25–40% of all prescription drugs contain active ingredients derived from plants in the United States alone. Many countries rely on these medicinal plants for the health and well being of its population, but the market demand has led to an increased pressure on the natural resources that lend to the production of some of these plants. The most serious proximate threats when extracting medicinal plants generally are habitat loss, habitat degradation, and over harvesting. Developing markets for natural products, particularly those that are harvested from the wild, can trigger a demand that cannot be met by available or legal supplies and demands a conservation initiative so the local populations are not exploited, causing more damage to their resources. Many times populations are taken advantage of for their resources and knowledge, which can often be for financial gain. Conservation of medicinal plants in its biocultural perspective not only implies conservation of biodiversity but also places an equal emphasis on conservation of cultural diversity.

Around the Globe

Asia represents one of the most important centers of knowledge with regard to the use of plant species for treatment of various diseases. Kunwar states, it has been estimated that the Himalayan region harbors over 10,000 species of medicinal and aromatic plants, supporting the livelihoods of about 600 million people living in the area. In Nepal they use a traditional healing system that is called Ayurveda, which is influenced by Buddhism and Hinduism's central ideas of balance in life. High altitude medicinal plants provide quality products, and this is the reason why they are often the first choice of local users as immediate therapy and by pharmaceutical companies as precious ingredients. When it comes to profits made in the communities up to 50% of the Nepal's rural household's income is derived from commercial collection of medicinal and aromatic plants.

KwaZulu-Natal, South Africa community of Mnoqobokazi has high unemployment rates in the area and reliance on subsistence agriculture and wild produce is still high. Socioeconomic factors such as low education levels and lack of access to western health care have been cited as important reasons for reliance on indigenous medicine in South Africa. Both villagers and healers in the area would cultivate one or more species because they could only be found far away, or were frequently used, or had to be fresh when used, or they were planted as protection against witchcraft. Also there were ten people from Mnqobokazi, mainly women, who harvested plants on a commercial basis to conserve. An interview with a conservation officer at the Wetland Park claimed that harvesting of medicinal plants was not a problem in this part as it was further north. In other parts of South Africa the most frequently used medicinal plants are slow-growing forest trees, in which the bark and underground parts are mainly the parts utilized. Because there is a high demand for such resources, the trees are becoming endangered and a lot of the collection is unrestricted. Regulations are now being placed on some of the resources that originally had been exploited and many schools and research facilities are working together to come up with new ways to foster their beloved trees and still manage to get what is needed from the trees as well by proposing the idea of substituting the bark or underground parts with leaves of the same plant.

Samoa has had a great influence on western medicine when it comes to finding a cure for HIV/AIDS. New research has shown that the isolation of prostratin, found in the bark of the Samoan mamala tree, from *Homalanthus nutans* has led to the extreme potency against HIV-1. Both the National Cancer Institute and Brigham Young University have guaranteed to return to the Samoan people a significant portion of any royalties. Paul Alan Cox, an American ethnobotanist, raised money based on awareness of environmental degradation due to logging, in order to protect the 30,000-acre (120 km^2) lowland forest of Falealupo village on the island of Savai'i. The Swedish Society for the Conservation of Nature established three new indigenously controlled preserves. Controlled preserves cause controversy because in traditional Polynesian societies, land, including the natural plant and animal populations, which occupied it, were viewed as sacred and an ancestral inheritance. Western approaches to conservation on indigenous land and within an indigenous community must collaborate and understand indigenous knowledge systems in order to conserve cultural identity. Paul Cox stated that, "the loss of these indigenous knowledge systems may yet prove to be one of the greatest tragedies of our age". The U.S National Park Service officials, the American Samoan Government, and the traditional chiefs *(matai)* and orators of the villages of Tafua had agreed to lease their lands for 50 years to the U.S National Park Service in order to protect American Samoa's rain forests. The Tafua rain forest received funding from the Swedish International Development Authority, which was used to secure water supply, improve roads, and used for assistance in the development of village-based environmental tourism in Tafua. Cox explains, "all parties to these agreements agree that any development of tourism must be village initiatives, rather than foreign initiatives, and must be carefully planned and controlled so that the Samoan culture in these areas is not jeopardized".

Bioprospecting

A white Rosy Periwinkle

Bioprospecting is the process of discovery and commercialization of new products based on biological resources. Despite indigenous knowledge being intuitively helpful, bioprospecting has only recently begun to incorporate such knowledge in focusing screening efforts for bioactive compounds. Bioprospecting may involve biopiracy, the exploitative appropriation of indigenous forms of knowledge by commercial actors, and also includes the search for previously unknown compounds in organisms that have never been used in traditional medicine before.

Biopiracy

Biopiracy describes a practice in which indigenous knowledge of nature, originating with indigenous peoples, is used by others for profit, without permission from and with little or no compensation or recognition to the indigenous people themselves. For example, when bioprospectors draw on indigenous knowledge of medicinal plants which is later patented by medical companies without recognizing the fact that the knowledge is not new, or invented by the patenter, and depriving the indigenous community to the rights to commercial exploitation of the technology that they themselves had developed. Critics of this practice, such as Greenpeace, claim these practices contribute to inequality between developing countries rich in biodiversity, and developed countries hosting companies that engage in 'biopiracy'.

Famous Cases

The Maya ICBG Controversy

The Maya ICBG bioprospecting controversy took place in 1999-2000, when the International Cooperative Biodiversity Group led by ethnobiologist Brent Berlin was accused of being engaged in unethical forms of bioprospecting by several NGOs and indigenous organizations. The ICBG aimed to document the biodiversity of Chiapas, Mexico and the ethnobotanical knowledge of the indigenous Maya people - in order to ascertain whether there were possibilities of developing medical products based on any of the plants used by the indigenous groups.

The Maya ICBG case was among the first to draw attention to the problems of distinguishing between benign forms of bioprospecting and unethical biopiracy, and to the difficulties of securing community participation and prior informed consent for would-be bioprospectors.

The Rosy Periwinkle

The rosy periwinkle case dates from the 1950s. The rosy periwinkle, while native to Madagascar, had been widely introduced into other tropical countries around the world well before the discovery of vincristine. This meant that researchers could obtain local knowledge from one country and plant samples from another. The use of the plant as a cure for diabetes was the original stimulus for research. Effectiveness in the treatment of both Hodgkin's Disease and leukemia were discovered instead. Different countries are reported as having acquired different beliefs about the medical properties of the plant. The Hodgkin's lymphoma chemotherapeutic drug vinblastine is derivable from the rosy periwinkle.

The Neem Tree

In 1994, the U.S. Department of Agriculture and WR Grace received a European patent on methods of controlling fungal infections in plants using a composition that included extracts from the neem tree (*Azadirachta indica*), which grows throughout India and Nepal. In 2000 the patent was successfully opposed by several groups from EU and India including the EU Green Party, Vandana Shiva, and the International Federation of Organic Agriculture Movements (IFOAM) on the basis that the fungicidal activity of neem extract had long been known in Indian traditional medicine. WR Grace appealed, and lost that appeal in 2005.

A neem tree

The Enola Bean

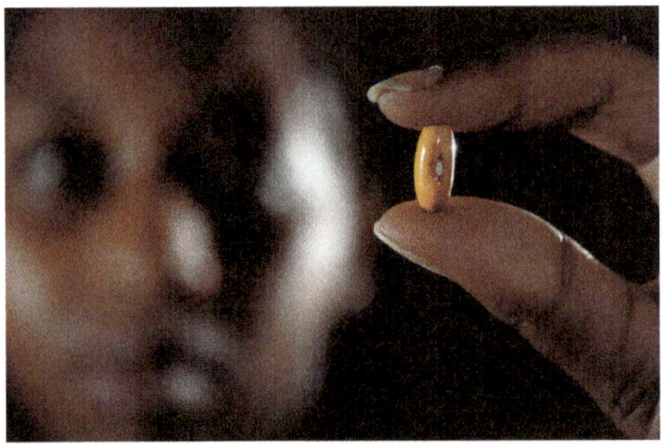

The Enola bean

The Enola bean is a variety of Mexican yellow bean, so called after the wife of the man who patented it in 1999. The allegedly distinguishing feature of the variety is seeds of a specific shade of yellow. The patent-holder subsequently sued a large number of importers of Mexican yellow beans with the following result: "...export sales immediately dropped over 90% among importers that had been selling these beans for years, causing economic damage to more than 22,000 farmers in northern Mexico who depended on sales of this bean." A lawsuit was filed on behalf of the farmers, and on April 14, 2005 the US-PTO ruled in favor of the farmers. An appeal was heard on 16 January 2008, and the patent was revoked in May 2008. An appeal to the court against the revocation was unsuccessful (Decided October 2009).

Basmati Rice

In 2000, the US corporation RiceTec (a subsidiary of RiceTec AG of Liechtenstein) attempted to patent certain hybrids of basmati rice and semidwarf long-grain rice. The Indian government intervened and several claims of the patent were invalidated. Meanwhile, the European Commission has agreed to protect basmati rice under its regulations pertaining to geographical indications.

Hoodia

Hoodia, a succulent plant, originates from the Kalahari Desert of South Africa. For generations it has been known to the traditionally living San people as an appetite suppressant. In 1996 South Africa's Council for Scientific and Industrial Research began working with companies, including Unilever, to develop dietary supplements based on hoodia. Originally the San people were not scheduled to receive any benefits from the commercialization of their traditional knowledge, but in 2003 the South African San Council made an agreement with CSIR in which they would receive from 6 to 8% of the revenue from the sale of Hoodia products.

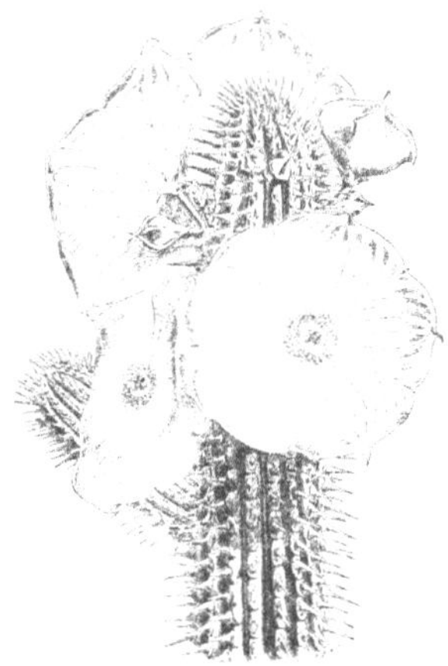
Gordon's Hoodia (*Hoodia Gordonii*). Glühende Pflanze.

The succulent *Hoodia*

In 2008 after having invested €20 million in R&D on hoodia as a potential ingredient in dietary supplements for weight loss, Unilever terminated the project because their clinical studies did not show that hoodia was safe and effective enough to bring to market.

Further Cases

The following is a selection of some of the further cases in recent biopiracy studies. Most of them do not relate to traditional medicines.

- A large selection of African biopiracy cases are discussed at GhanaWeb's *Public Agenda* (March 31, 2006)
- The case of the Maya people's *pozol* drink is discussed in *The New England Journal of International and Comparative Law* and at Global Exchange
- The case of the Maya and other people's use of *Mimosa tenuifolia*, including many other such cases in general are discussed at GRAIN
- The case of the Andean *maca* radish is discussed in the American University's Trade Environment Database

- The *United Kingdom Select Committee on Environmental Audit 1999; Appendices to the Minutes of Evidence, Appendix 7: Trade Related Intellectual Property Rights (TRIPs) and Farmers' Rights* interestingly lists and describes the cases of turmeric (India), karela (India), quinoa (Bolivia), brazzein berries (Gabon), and others.
- Captopril

Legal and Political Aspects

Patent Law

One common misunderstanding is that pharmaceutical companies patent the plants they collect. While obtaining a patent on a naturally occurring organism as previously known or used is not possible, patents may be taken out on specific chemicals isolated or developed from plants. Often these patents are obtained with a stated and researched use of those chemicals. Generally the existence, structure and synthesis of those compounds is not a part of the indigenous medical knowledge that led researchers to analyze the plant in the first place. As a result, even if the indigenous medical knowledge is taken as prior art, that knowledge does not by itself make the active chemical compound "obvious," which is the standard applied under patent law.

In the United States, patent law can be used to protect "isolated and purified" compounds - even, in one instance, a new chemical element (USP 3,156,523). In 1873, Louis Pasteur patented a "yeast" which was "free from disease" (patent #141072). Patents covering biological inventions have been treated similarly. In the 1980 case of *Diamond v. Chakrabarty*, the Supreme Court upheld a patent on a bacterium that had been genetically modified to consume petroleum, reasoning that U.S. law permits patents on "anything under the sun that is made by man." The United States Patent and Trademark Office (USPTO) has observed that "a patent on a gene covers the isolated and purified gene but does not cover the gene as it occurs in nature".

Also possible under US law is patenting a cultivar, a new variety of an existing organism. The patent on the enola bean (now revoked) was an example of this sort of patent. The intellectual property laws of the US also recognize plant breeders' rights under the Plant Variety Protection Act, 7 U.S.C. 2321-2582.

Convention on Biological Diversity (CBD)

The CBD came into force in 1993. It secured rights to control access to genetic resources for the countries in which those resources are located. One objective of the CBD is to enable lesser-developed countries to better benefit from their resources and traditional knowledge. Under the rules of the CBD, bioprospectors are required to obtain informed consent to access such resources, and must share any benefits with the biodiversity-rich country. However, some critics believe that the CBD has failed to establish appropriate regulations to prevent biopiracy. Others claim that the main problem is the failure of national governments to pass appropriate laws implementing the provisions of the CBD. The Nagoya Protocol to the CBD (negotiated in 2010, expected to come into force in 2014) will provide further regulations. The CBD has been ratified by all countries in the world except for Andorra, Holy See and United States. The 1994 Agreement on Trade-Related Aspects of Intellectual Property Rights (TRIPs) and the 2001 International

Treaty on Plant Genetic Resources for Food and Agriculture are further relevant international agreements.

Bioprospecting Contracts

The ethical debate has sparked a new branch of international patent and trade law. Bioprospecting contracts lay down the rules, between researchers and countries, of benefit sharing and can bring royalties to lesser-developed countries. However, the fairness of these contracts has been a subject of debate. Unethical bioprospecting contracts (as distinct from ethical ones) can be viewed as a new form of biopiracy.

An extensively discussed example of a bioprospecting contract is the agreement between Merck and INBio of Costa Rica.

On June 14, 2011, Colombia approved a policy for the sustainable commercial use of its biodiversity resources, primarily through the development of biotechnology research. It includes plans to set up a national company for bioprospecting to link up with the commercial sector and will be backed with US$14 million in government funds over the next four years.

Traditional Knowledge Database

In response to concerns of biopiracy raised by research into turmeric, neem and basmati rice, the Government of India has been translating and publishing ancient manuscripts containing old remedies in electronic form, and in 2001 the Traditional Knowledge Digital Library was set up as a repository of 1200 formulations of various systems of Indian medicine, such as Ayurveda, Unani and Siddha. The texts are being recorded from Tamil, Sanskrit, Urdu, Persian and Arabic; made available to patent offices in English, German, French, Japanese and Spanish. The aim is to protect India's heritage from being exploited by foreign companies. Hundreds of yoga poses are also kept in the collection. The project has been criticized by a spokesman for the pharmaceutical industry as "a solution in search of a problem". The library has also signed agreements with leading international patent offices such as European Patent Office (EPO), United Kingdom Trademark & Patent Office (UKTPO) and the United States Patent and Trademark Office to protect traditional knowledge from biopiracy as it allows patent examiners at International Patent Offices to access TKDL databases for patent search and examination purposes. The database is not available to the general public, but the articles to which it refers are (though typically in a not widely understood language such as Sanskrit). In this way the database prevents subsequent patenting without making the information to which it refers readily available for public use

References

- Schools History Project (26 September 1996). Medicine & Health Through Time: an SHP Development Study. Hodder Education. ISBN 978-0719552656
- Jonathan D. Eldredge (2003), "The Randomised Controlled Trial design: unrecognized opportunities for health sciences librarianship", Health Information and Libraries Journal 20, p. 34–44
- Ortiz de Montellano, B (1975). "Empirical Aztec medicine". Science. 188 (4185): 215–20. PMID 1090996. doi:10.1126/science.1090996

- Goldberg, Danielle (2003). "Jack and the Enola Bean". TED Case Studies Number xxx. Danielle Goldberg. Retrieved 2013-11-04
- Moerman, Daniel E. (2009). Native American Medicinal Plants: an ethnobotanical dictionary. Portland, OR / London: Timber Press. ISBN 978-0-88192-987-4
- Soltis, D. E.; Soltis, P. S. (2004). "Amborella not a "basal angiosperm"? Not so fast". American Journal of Botany. 91 (6): 997–1001. PMID 21653455. doi:10.3732/ajb.91.6.997
- Rennard, B. O.; Ertl, RF; Gossman, GL; Robbins, RA; Rennard, SI (2000). "Chicken Soup Inhibits Neutrophil Chemotaxis in Vitro". Chest. 118 (4): 1150–7. PMID 11035691. doi:10.1378/chest.118.4.1150
- "As easy as APG III - Scientists revise the system of classifying flowering plants". The Linnean Society of London. 2009-10-08. Retrieved 2009-10-02
- Fienup-Riordan, Ann (1994). Boundaries and Passages: Rule and Ritual in Yup'ik Eskimo Oral Tradition. Norman, OK: University of Oklahoma Press. ISBN 0-585-12190-7
- Cox, PA (1993). "Saving the ethnopharmacological heritage of Samoa". Journal of Ethnopharmacology. 38 (2–3): 181–8. PMID 8510467. doi:10.1016/0378-8741(93)90014-v
- Christenhusz, M. J. M.; Byng, J. W. (2016). "The number of known plants species in the world and its annual increase". Phytotaxa. Magnolia Press. 261 (3): 201–217. doi:10.11646/phytotaxa.261.3.1
- Girish Dwivedi, Shridhar Dwivedi (2007). History of Medicine: Sushruta – the Clinician – Teacher par Excellence (PDF). National Informatics Centre. Retrieved 2008-10-08
- Ackerknecht, Erwin Heinz (1982) [1955]. A Short History of Medicine (Johns Hopkins Paperbacks ed.). Baltimore: Johns Hopkins University Press. ISBN 0-8018-2726-4
- Koshy, Jacob P.; gen patent claim (2010-04-28). "CSIR wing objects to Avesth". live mint & The Wall Street Journal. HT Media. Retrieved 2013-11-04
- L. Pallottini; E. Garcia; J. Kami; G. Barcaccia; P. Gepts (1 May 2004). "The Genetic Anatomy of a Patented Yellow Bean". Crop Science. 44 (3): 968–977. doi:10.2135/cropsci2004.0968
- Wilford, John Noble (December 8, 1998). "Lessons in Iceman's Prehistoric Medicine Kit". The New York Times. New York. Retrieved December 30, 2015
- Slikkerveer, L. J. (1990). Plural medical systems in the Horn of Africa: the legacy of "Sheikh" Hippocrates. London: Kegan Paul International. ISBN 0-7103-0203-7
- Ewen Callaway (December 2013). "Shrub genome reveals secrets of flower power". Nature. doi:10.1038/nature.2013.14426
- "Babylon to Birmingham, A short journey through medicine to the end of the 18th Century". Revolutionary Players. History West Midlands. Retrieved 2015-12-30
- Fahd, Toufic (1996). "Botany and agriculture". In Rashed, Roshdi; Morelon, Régis. Encyclopedia of the History of Arabic Science. Routledge. p. 815. ISBN 0-415-02063-8
- Eberlee, John (2000-01-21). "Assessing the Benefits of Bioprospecting in Latin America" (PDF). IDRC Reports Online. IDRC. Retrieved 2013-11-03

2

Various Types of Medicational Plants

Hebe stricta is used to control diarrhea and dysentery whereas chamomile can cure inflammation and menstrual disorders. Laurelia novae-zelandiae, leptospermum scoparium, macropiper excelsum, celery and ballota nigra. The chapter offers an insightful focus, keeping in mind the complex subject matter.

Hebe Stricta

Hebe stricta, commonly called koromiko, is a plant of the family Plantaginaceae, which is endemic to New Zealand. Recently Hebe stricta has been reclassified and the botanical name has changed to *Veronica stricta*.

A hebe found throughout the North Island with long pale green leaves and tiny white flowers in summer into a dense inflourences longer than the leaves. The *koromiko* is a hardy plant and does not tolerate shade. Full sun only and prefers open habitats on edge of bush and wetlands as it tolerates wet areas.

Varieties

There are two varieties of this species:

- *Hebe stricta* var. *atkinsonii*
- *Hebe stricta* var. *macroura*

Uses

The young leaves and shoots is eaten to control Diarrhoea and dysentery It is a pioneer plant useful to plant as a nurse crop for revegetation planting.

Synonyms

These synonyms are cited by Germplasm Resources Information Network

- *Hebe macroura* (Hook. f. ex Benth.) Cockayne & Allan [≡ *Hebe stricta* var. *macroura*]
- *Hebe parkinsoniana* (Colenso) Cockayne
- *Hebe salicifolia* var. *angustissima* (Cockayne) Cockayne & Allan [= *Hebe stricta* var. *atkinsonii*]

- *Hebe salicifolia* var. *atkinsonii* (Cockayne) Cockayne & Allan [≡ *Hebe stricta* var. *atkinsonii*]
- *Hebe salicifolia* var. *stricta* (Banks & Sol. ex Benth.) Cockayne & Allan
- *Veronica macroura* Hook. f. ex Benth. [≡ *Hebe stricta* var. *macroura*]
- *Veronica parkinsoniana* Colenso
- *Veronica salicifolia* var. *angustissima* Cockayne [= Hebe stricta var. atkinsonii]
- *Veronica salicifolia* var. *atkinsonii Cockayne* [≡ *Hebe stricta* var. *atkinsonii*]
- *Veronica salicifolia* var. *stricta* (Banks & Sol. ex Benth.) Hook. f.
- *Veronica stricta* Banks & Sol. ex Benth. (basionym)

Laurelia Novae-zelandiae

Laurelia novae-zelandiae, also called pukatea, is a large evergreen tree, endemic to the forests of New Zealand. Pukatea has 'toothed' leaves and produces small flowers. It is a species in the Atherospermataceae (formerly Monimiaceae) family, typical representative of laurel forest ecoregion.

Distribution

Pukatea is generally found in lowland forest and grows throughout the North Island of New Zealand, and the northern third of the South Island, usually where moisture is plentiful, such as in damp, lowland forests, gullies, and on the edges of streams.

Pukatea grows well in poorly drained soil, but is equally at home on hillsides. It requires a temperate to warm subtropical climate, but also a frost-free environment with only very slight winter frosts not below 24.8°F (-4°C), and with high summer heat. Growth is best on well-drained, slightly acidic soils rich in organic matter.

Description

Leaves of a pukatea seedling

The buttressed trunk of a pukatea

Pukatea grows slowly to a height of 130 feet (40 m), usually 115 feet (35 m), and is the only New Zealand native tree developing large plank buttresses to support the tree's growth in swamp or shallow-soil areas. *L. novae-zelandiae* has specialized respiratory root structures called pneumatophores in certain waterlogged ground or mud. These fragrant trees are characteristic of the lower strata of the tropical rainforest. The tree has thin bark and a pale brownish-grey trunk that becomes attractively buttressed at the base. Its dark green, glossy, elliptical leaves are 2-3 inches (5-7 cm) long and have coarsely serrated edges and paler undersides. The odorous opposite leaves have oil cells in the parenchyma, and brochidodromous venations. Juvenile leaves and stems can be difficult to distinguish from another native tree hutu, to which it is only distantly related. Pukatea is a mostly dioecious species, male and female flowers are on separate individuals. Some specimens had a ratio as high as 100 male flowers to every female or hermaphrodite one. These results suggest that the species is not truly dioecious. The female and hermaphrodite flowers are very similar. The flowers are tiny, inconspicuous, and in small racemes. The star-shaped flowers are whitish with yellow glistening glands and scarlet anther flaps. The glands at the base of the stamens in *L. novae-zelandiae* secrete nectar that accumulates at the base of the flower. This attracts a large number of bees, blowflies, small flightless animals, and Bombyliidae to the flowers. The nectar is visible as a colourless liquid which has a glistening appearance to the outer faces of the glands. Nectar runs down from the glands and accumulates on the floor of the flower as a result of continuing secretion. The fruit are little pear-shaped capsules which contain numerous achenes attached to fine feathery strands which aid their dispersal by the wind. Often, only a few seeds are viable.

Uses

Historically, the light, but tough timber of pukatea has been used for boat building. The timber was used by Maori to create figureheads for canoes. It is rather soft, but very strong. It yields a pale hardwood that is difficult to split and that upon impact rather than to break, it dents. Its wood is pale-yellowish, with growth rings, and is homogeneous and fine-textured. An extract from the bark containing the alkaloid pukateine is used in traditional Māori herbal medicine as an analgesic. The pulp of the cambium was boiled in water and the resulting liquid used for treating tuberculosis.

Chamomile

German chamomile

Roman chamomile

Chamomile or camomile is the common name for several daisy-like plants of the family Asteraceae that are commonly used to make herb infusions to serve various medicinal purposes. Popular uses of chamomile preparations include treating hay fever, inflammation, muscle spasms, menstrual disorders, insomnia, ulcers, gastrointestinal disorders, and hemorrhoids. Chamomile tea is also used to treat skin conditions such as eczema, chickenpox, and psoriasis.

Etymology

The word "chamomile" derives, via French and Latin, from Greek (*khamaimēlon*), i.e. "earth apple", from (*khamai*) "on the ground" and (*mēlon*) "apple". The more common British spelling "camomile," is the older one in English, while the spelling "chamomile" corresponds to the Latin and Greek source. The spelling *camomile* is a derivation from the French.

Species

Loose leaf chamomile tea

Some commonly used species include:

- *Matricaria chamomilla* (also known as *Water of Youth'), German chamomile or wild chamomile, the most commonly used species*

- *Chamaemelum nobile*, Roman, English or garden chamomile, also frequently used, (*C. nobile 'Treneague'* is normally used to create a chamomile lawn).

A number of other species' common names include the word "chamomile". This does not mean they are used in the same manner as the species used in the herbal tea known as "chamomile." Plants including the common name "chamomile," of the family Asteraceae, are:

- *Anthemis arvensis*, corn, scentless or field chamomile
- *Anthemis cotula*, stinking chamomile
- *Cladanthus mixtus*, Moroccan chamomile
- *Cota tinctoria*, dyer's, golden, oxeye, or yellow chamomile
- *Eriocephalus punctulatus*, Cape chamomile
- *Matricaria discoidea*, wild chamomile or pineapple weed
- *Tripleurospermum inodorum*, wild, scentless or false chamomile

Tea / Herbal Infusion

Chamomile 'tea' is an herbal infusion made from dried chamomile flowers and hot water, though does not contain black, green, yellow or white tea (Camellia sinensis). There are two main types of chamomile that are used to produce the infusion; German chamomile (*Matricaria recutita*) and Roman chamomile (*Chamaemelum nobile*). Chamomile 'tea' is said to be a digestive relaxant.

Medical use

German Chamomile has been used for treatment of inflammation associated with hemorrhoids when topically applied. There is level B evidence that German chamomile possesses anti-anxiety properties and could be used to treat stress and insomnia.

Pharmacology

Major chemical compounds present within chamomile include apigenin and alpha-bisabolol. Other compounds in chamomile include: sesquiterpenes, terpenoids, flavonoids, coumarins such as herniarin and umbelliferone, phenylpropanoids such as chlorogenic acid and caffeic acid, flavones such as apigenin and luteolin, flavanols such as quercetin and rutin, and polyacetylenes. Apigenin has demonstrated chemopreventive effects against cancer cells in the laboratory, and alpha-bisabolol has been shown to have antiseptic properties, anti-inflammatory properties, and reduces pepsin secretion without altering secretion of stomach acid.

Anticancer effect – Studies have shown that chamomile extracts have *in vitro* growth inhibitory effects on cancer cells in skin, prostate, breast, ovarian, prostate cancer cell lines with minimal effects on normal cells.

Anticoagulant effect – Coumarin compounds in chamomile such as herniarin and umbelliferone may have blood-thinning properties. However, the mechanism is not well understood.

Anti-inflammatory effect – Several chemical constituents of chamomile such as bisabolol, chamazulene, apigenin, and luteolin possess anti-inflammatory properties although the exact mechanism is not well characterized.

Antispasmodic/antidiarrheal effects – Bisabolol and flavonoids have demonstrated antispasmodic effects in animal experiments. In human studies, chamomile tea in combination with other herbs (vervain, licorice, fennel, balm mint) was shown to be effective in treating colic in children. Flavonoids and coumarins are considered smooth muscle relaxants.

CNS/sensory effects - Chemical compounds present within chamomile bind to GABA receptors, modulate monoamine neurotransmission, and have neuroendocrine effects.

Drug Interactions

Apigenin and other compounds may interact with medications causing drug-drug interactions, some of the possible interactions include those with antiplatelet agents, anticoagulant agents, and nonsteroidal anti-inflammatory agents. Apigenin was found to interact with antiarrhythmic agents and antihypertensive agents in animal research. Other interactions include those against sedative agents, antibiotic agents, and antianxiety agents. Remarkable symptoms are exacerbation of effects of these agents that are used in combination with chamomile.

While chamomile exhibits some anti-inflammatory effects by itself, it is not recommended that it be taken concurrently with Aspirin or non-salicylate NSAIDs (non-steroidal anti-inflammatory drugs) as it is unknown if a clinically significant herb-drug interaction exists.

"Chamomile consists of several ingredients including coumarin, glycoside, heniarin, flavonoid, farnesol, nerolidol and germacranolide. Despite the presence of coumarin, as chamomiles effect on the coagulation system has not yet been studied, it is unknown if a clinically significant drug-herb interaction exists with antiplatelet/anticoagulant drugs. However, until more information is available, it is not recommended to use these substances concurrently."

Adverse Reactions

People who are allergic to ragweed (also in the daisy family) may also be allergic to chamomile, due to cross-reactivity. However, there is still some debate as to whether people with reported allergies to chamomile were actually exposed to chamomile and not a plant of similar appearance.

Pregnancy / Lactation

Because chamomile has been known to cause uterine contractions that can invoke miscarriage, the U.S. National Institutes of Health recommends that pregnant and nursing mothers not consume Roman chamomile (*Chamaemelum nobile*).

Cosmetic Applications

Chamomile is frequently added to skin cosmetics to serve as an emollient, and for its anti-inflammatory effects. Chamomile is also often used to enhance the color of blonde hair.

Agriculture

The chamomile plant is known to be susceptible to many fungi, insects, and viruses. Fungi such as *Albugo tragopogonis* (white rust), *Cylindrosporium matricariae*, *Erysiphe cichoracearum* (powdery mildew), and *Sphaerotheca macularis* (powdery mildew) are known pathogens of the chamomile plant. Aphids have been observed feeding on chamomile plants and the moth *Autographa chryson* causes defoliation.

Oil

German chamomile oil is used as a diffuser for aromatherapy benefits; and is also used to treat wounds and be blended with other essential oils such as lavender and rose.

Research

Chemical components of chamomile extract have demonstrated anti-inflammatory, antihyperglycemic, and anticancer properties when examined in test tubes and in animal studies.

Leptospermum Scoparium

Leptospermum scoparium, commonly called mānuka, manuka myrtle, New Zealand teatree, broom tea-tree, or just tea tree, is a species of flowering plant in the myrtle family Myrtaceae, native to Australia and New Zealand.

Description

It is a prolific scrub-type tree and is often one of the first species to regenerate on cleared land. It is typically a shrub growing to 2–5 m (7–16 ft) tall, but can grow into a moderately sized tree, up to 15 m (49 ft) or so in height. It is evergreen, with dense branching and small leaves 7–20 mm long and 2–6 mm broad, with a short spine tip. The flowers are white, occasionally pink, 8–15 mm (rarely up to 25 mm) in diameter, with five petals. This species is often confused with the closely related species kānuka – the easiest way to tell the difference between the two species in the field is to feel their foliage – mānuka leaves are prickly, while kānuka leaves are soft. The wood is tough and hard.

History and Habitat

Evidence suggests that *L. scoparium* originated in Australia before the onset of the Miocene aridity, and dispersed relatively recently from eastern Australia to New Zealand. It has been postulated that on arrival in New Zealand, *L. scoparium* became established in limited edaphically suitable areas until the arrival of the Polynesian people, whose fire and forest-clearing brought about the low-nutrient-status soils for which it was preadapted in its homeland. It is now found throughout New Zealand, but is particularly common on the drier east coasts of the North and South Islands, and in Australia in Tasmania, Victoria and New South Wales.

Ecology

Pests and Diseases

The adult manuka beetle and its larvae feed on *L. scoparium.*

Names

Tea tree is the common name. Mānuka (from Māori) is also commonly used. "Jelly bush" is also used in Australia. Tea tree arose because Captain Cook used the leaves to make a 'tea' drink.

Cultivation

Numerous cultivars have been developed for garden use, of which the following have gained the Royal Horticultural Society's Award of Garden Merit:-

- 'Kiwi'
- 'Red Damask'
- 'Nichollsii Nanum'
- 'Silver Sheen'

Many more cultivars are available in New Zealand but often the plant performs better in cultivation overseas. This is because in its homeland it is subject to attack by scale insects that secrete a honeydew on which grows a sooty mold that eventually debilitates the plant. Because of this, attempts have been made, with limited commercial success, to cross the showy New Zealand cultivars with mold-resistant Australian *Leptospermum* species.

Uses

Manuka (*Leptospermum scoparium*) essential oil in a clear glass vial

The wood was often used for tool handles. Mānuka sawdust imparts a delicious flavour when used for smoking meats and fish. It is cultivated in New Zealand for mānuka honey, produced when honeybees gather the nectar from its flowers, and for the pharmaceutical industry. It is also used for carving. An essential oil, for which many medicinal claims are made, is produced by steam distillation of its leaves. Manuka, as it is called by most New Zealanders, was used in pre-European times by the traditional people of New Zealand, the Maori, and still is.

A decoction of the leaves was drunk for urinary complaints and as a febrifuge (an agent for reducing fever). The steam from leaves boiled in water was inhaled for head colds. A decoction was prepared from the leaves and bark and the warm liquid was rubbed on stiff muscles and aching joints. The emollient white gum, called pai Manuka, was given to nursing babies and also used to treat scalds and burns. Chewing the bark is said to have a relaxing effect and it enhances sleep.

Parakeets and Parasites

Kakariki parakeets (*Cyanoramphus*) use the leaves and bark of mānuka and kānuka to rid themselves of parasites. Apart from ingesting the material, they also chew it, mix it with preen gland oil and apply it to their feathers.

Macropiper Excelsum

Piper excelsum (formerly known as Macropiper excelsum), called Kawakawa, is a small tree of which the subspecies *M. excelsum* subsp. *excelsum* is endemic to New Zealand; the subspecies *M. e.* subsp. *psittacorum* is found on Lord Howe Island, Norfolk Island and the Kermadec Islands

Description

It is found throughout the North Island, and as far south as Okarito (43.20 °S) on the West Coast and Banks Peninsula (43.5 °S) on the east coast of the South Island. The leaves are often covered with insect holes. The images depict the variety *majus* which has larger and more glossy leaves than *P. excelsum*. The name *Kawakawa* in Māori refers to the bitter taste of the leaves, from *kawa* bitter.

Leaves

The Kawakawa leaves are about 5–10 cm long by 6–12 cm wide; they are opposite to each other, broadly rounded with a short drawn-out tip and are heart-shaped at their bases. The leaves are deep green in color if growing in the forest but may be yellowish-green when growing in more open situations.

Flowers

The flowers are produced on greenish, erect spikes that are 2.5-7.5 cm long. Kawakawa flowers are quite minute and very closely placed around the spike. After pollination the flowers gradually swell and become fleshy to form small, berry-like fruits that are yellow to bright orange.

Berries

Each berry cluster is the size of a small finger. Ripening period is January and February. These fruits are favoured by kereru, or New Zealand pigeon (*Hemiphaga novaeseelandiae*) and tui (*Prosthemadera novaeseelandiae*).

Uses

Kawakawa leaves and fruiting spikes

Kawakawa is a traditional medicinal plant of the Māori. An infusion is made from the leaves or roots, and used for bladder problems, boils, bruises, to relieve pain or toothache, or as a general tonic. The sweet edible yellow berries (most often found in summer on female trees) of the plant were eaten as a diuretic.

It also is important in cultural contexts: host people of a marae wave leaves of kawakawa to welcome guests. At a tangi, both hosts and guests may wear wreaths of kawakawa on the head as a sign of mourning.

It is now available as an ornamental plant for gardens.

Relationship with kava

Kawakawa is sometimes called "Maori kava" and is often confused with the kava plant (*Piper methysticum*). While the two plants look similar and have similar names, they are different, albeit related, species.

Kava is a traditional plant and beverage of the South Pacific. The roots of the plant are used to produce a drink with medicinal, sedative, anesthetic, euphoriant, and entheogenic properties. It is most likely not a coincidence that this plant has a similar name to kawakawa. One source stated: "In New Zealand, where the climate is too cold for kava, the Maori gave the name kawa-kawa to another Piperaceae, M. excelsum, in memory of the kava plants they undoubtedly brought with them and unsuccessfully attempted to cultivate. The Maori word kawa also means "ceremonial protocol", recalling the stylized consumption of the drug typical of Polynesian societies."

Ballota Nigra

Ballota nigra (black horehound) is a perennial herb of the family Lamiaceae. It is native to the Mediterranean region and to central Asia, and it can be found throughout Europe. It is also naturalized in Argentina, New Zealand, and the Eastern United States. It blooms in the Northern Hemisphere from May to August.

Description

Ballota nigra has a very strong characteristic smell that reminds of mould or humidity, and can be recognised by its clusters of hairy, reddish-purple flowers. It can grow up to 3 feet in height.

Morphology

Stem and Root

It has herbaceous ascending stems, wooden and branched at bottom, covered by down folded hairs. The plant has a taproot system.

Leaves

Leaves are opposite and decussate, and range from oval-lanceolate to heart-shaped, with crenate or dentate border. Leaves, dark green and usually pubescent, measure 3–8 cm per 2–6 cm, and have 1–3 cm petiole. Upper face is wrinkled, with a net-like vein pattern.

Flowers

Flowers are organized in verticillasters, subspherical to about one-sided, with 15 to 30 flowers. Each verticillaster consist of two condensed dichasial cymes at axils of normal leaves.

Flower has an actinomorphic calyx (length 9–10 mm, width 7 mm), made up by five sepals fused together in a tube with five teeths; and a labiate corolla of 12–13 mm, ranging from pink to pale purple to withish. The corolla consist of a tube of about 6 mm and two lips; the upper one slightly concave (like a hood) and externally hairy; the lower one glabrous, with two minor lateral lobes and a major central bifid lobe. There are four didynamous stamens, running parallel under the upper lip, with glabrous filaments and yellow anthers. Ovary is superior, with a single white style and a 2-parted stigma.

Below the calyx there are five filiform bracts, 8 mm long.

Fruit

Each fertilized flower produces a tetrad of black nutlets, cylindrical to ovoid, 2 mm long, partially or fully covered by the calyx. The basal end is flat and attached to the receptacle, while the top end is rounded or pointed.

Biochemistry

Ballota nigra contains diterpenoids like marrubiin, ballonigrin, ballotinone, ballotenol and 7-acetoxymarrubiin. Also, it contains phenylpropanoids that have shown to be antioxidants.

Taxonomy and Etymology

The plant was described by Linnaeus in *Species Plantarum* (May 1753). The name *Ballota* comes from the Greek *ballo* (to reject), because of the strong offensive odor of the plant; cattle will not eat it. The specific name *nigra* could refer to the black colour of dried leaves.

The common name comes from the Old English words *har*, meaning "downy or hoary", and *hune*, meaning the plant itself. This name refers to the hairs that give the herb its distinctive appearance. In modern times, alternative medicine practitioners have referred to the plant as "seed of Horus" and suggested that horehound takes its name from Horus, the Egyptian sun god.

Distribution and Habitat

Ballota nigra is a nitrophilous plant; it grows in ruins, fallows and hedges, up to 1300 m. It prefers loose, calcareous (alkaline) soils. It tolerates temperatures as low as -5°/-10 °C.

Uses

Usually, the plant is used dry, harvested when blooming. Syrups can be made from fresh plants.

Subspecies

Recognized subspecies:

1. *Ballota nigra* subsp. *anatolica* P.H.Davis - Iran, Turkey
2. *Ballota nigra* subsp. *anomala* Greuter - Greece
3. *Ballota nigra* subsp. *foetida* (Vis.) Hayek - central + southern Europe; naturalized in Sweden, Ukraine, Cyprus, Turkey, Argentina
4. *Ballota nigra* subsp. kurdica P.H.Davis - Iran, Iraq, Turkey
5. *Ballota nigra* subsp. *nigra* - southern Europe, Great Britain, Sweden, Caucasus, Iran, Turkey; naturalized in Belgium, New Zealand, Argentina
6. *Ballota nigra* subsp. *ruderalis* (Sw.) Briq. - Mediterranean region; Canary Islands, Madeira, Azores
7. *Ballota nigra* subsp. *sericea* (Vandas) Patzak - Albania, Macedonia, Greece
8. *Ballota nigra* subsp. *velutina* (Posp.) Patzak - Slovenia, Croatia; naturalized in Argentina

Celery

Celery (*Apium graveolens*) is a marshland plant in the family Apiaceae that has been cultivated as a vegetable since antiquity. Celery has a long fibrous stalk tapering into leaves. Depending on location and cultivar, either its stalks, leaves, or hypocotyl are eaten and used in cooking.

Celery seed is also used as a spice; its extracts are used in medicines.

Description

Celery leaves are pinnate to bipinnate with rhombic leaflets 3–6 cm (1.2–2.4 in) long and 2–4 cm broad. The flowers are creamy-white, 2–3 mm in diameter, and are produced in dense compound umbels. The seeds are broad ovoid to globose, 1.5–2 mm long and wide. Modern cultivars have been selected for solid petioles, leaf stalks. A celery stalk readily separates into "strings" which are bundles of angular collenchyma cells exterior to the vascular bundles.

Head of celery, sold as a vegetable. Usually only the leaf stalks are eaten

Wild celery, *Apium graveolens* var. *graveolens*, grows to 1 m (3.3 ft) tall. It occurs around the globe. The first cultivation is thought to have happened in the mediterranean where the natural habitat were salty and wet or marshy soils near the coast and celery grew in agropyro-rumicion-plant communities.

North of the alps wild celery is found only in the foothill zone on soils with some salt content. It prefers moist or wet, nutrient rich, muddy soils. It cannot be found in Austria and is increasingly rare in Germany.

Etymology

First attested in English in 1664, the word "celery" derives from the French *céleri*, in turn from Italian *seleri*, the plural of *selero*, which comes from Late Latin *selinon*, the latinisation of the Greek (*selinon*), "celery". The earliest attested form of the word is the Mycenaean Greek *se-ri-no*, written in Linear B syllabic script.

Taxonomy

Cross-section of a 'Pascal' celery rib, the petiole

Celery was described by Carl Linnaeus in Volume One of his *Species Plantarum* in 1753.

Cultivation

The plants are raised from seed, sown either in a hot bed or in the open garden according to the season of the year, and, after one or two thinnings and transplantings, they are, on attaining a height of 15–20 cm (5.9–7.9 in), planted out in deep trenches for convenience of blanching, which is effected by earthing up to exclude light from the stems.

In the past, celery was grown as a vegetable for winter and early spring; it was perceived as a cleansing tonic, welcomed to counter the salt-sickness of a winter diet without greens based on salted meats. By the 19th century, the season for celery had been extended, to last from the beginning of September to late in April.

North America

In North America, commercial production of celery is dominated by the cultivar called 'Pascal' celery. Gardeners can grow a range of cultivars, many of which differ from the wild species, mainly in having stouter leaf stems. They are ranged under two classes, white and red. The stalks grow in tight, straight, parallel bunches, and are typically marketed fresh that way, without roots and just a little green leaf remaining.

The stalks are eaten raw, or as an ingredient in salads, or as a flavoring in soups, stews, and pot roasts.

Europe

In Europe, another popular variety is celeriac (also known as *celery root*), *Apium graveolens* var. *rapaceum*, grown because its hypocotyl forms a large bulb, white on the inside. The bulb could be kept for months in winter and mostly serves as a main ingredient in soup. It can also be ground up and used in salads. The leaves are used as seasoning; the small, fibrous stalks find only marginal use.

Asia

Leaf celery or Chinese celery, *Apium graveolens var. secalinum*, is a cultivar from East Asia that grows in marshlands.

Leaf celery, also known as Chinese celery

Leaf celery is most likely the oldest cultivated form of celery. Leaf celery has a characteristically thin skin stalks and a stronger taste and smell compared to other cultivars. It is used as a flavoring in soups and sometimes pickled as a side dish.

Wild

The wild form of celery is known as "smallage". It has a furrowed stalk with wedge-shaped leaves, the whole plant having a coarse, earthy taste, and a distinctive smell. The stalks are not usually eaten (except in soups or stews in French cuisine), but the leaves may be used in salads, and its seeds are those sold as a spice. With cultivation and blanching, the stalks lose their acidic qualities and assume the mild, sweetish, aromatic taste particular to celery as a salad plant.

Because wild celery is rarely eaten, yet susceptible to the same diseases as more well-used cultivars, it is often removed from fields to help prevent transmission of viruses like celery mosaic virus.

Harvesting and Storage

Harvesting occurs when the average size of celery in a field is marketable; due to extremely uniform crop growth, fields are harvested only once. The petioles and leaves are removed and harvested; celery is packed by size and quality (determined by colour, shape, straightness and thickness of petiole, stalk and midrib length and absence of disease, cracks, splits, insect damage and rot). During commercial harvesting, celery is packaged into cartons which contain between 36 and 48 stalks and weigh up to 60 pounds. Under optimal conditions, celery can be stored for up to seven weeks between 0 to 2 °C (32 to 36 °F). Inner stalks may continue growing if kept at temperatures above 0 °C (32 °F). Shelf life can be extended by packaging celery in anti-fogging, micro-perforated shrink wrap. Freshly cut petioles of celery are prone to decay, which can be prevented or reduced through the use of sharp blades during processing, gentle handling, and proper sanitation.

Sulfites

In the past, restaurants used to store celery in a container of water with powdered vegetable preservative, but it was found that the sulfites in the preservative caused allergic reactions in some people. In 1986, the U.S. Food and Drug Administration banned the use of sulfites on fruits and vegetables intended to be eaten raw.

Uses

Celery seed (*Apium graveolens*) essential oil

Celery is eaten around the world as a vegetable. In North America the crisp petiole (leaf stalk) is used. In Europe the hypocotyl is used as a root vegetable. The leaves are strongly flavoured and are used less often, either as a flavouring in soups and stews or as a dried herb. Celery, onions, and bell peppers are the "holy trinity" of Louisiana Creole and Cajun cuisine. Celery, onions, and carrots make up the French mirepoix, often used as a base for sauces and soups. Celery is a staple in many soups, such as chicken noodle soup.

Seeds

In temperate countries, celery is also grown for its seeds. Actually very small fruit, these "seeds" yield a valuable volatile oil used in the perfume and pharmaceutical industries. They contain an organic compound called apiole. Celery seeds can be used as flavouring or spice, either as whole seeds or ground.

Celery salt

The seeds can be ground and mixed with salt, to produce celery salt. Celery salt can be made from an extract of the roots or using dried leaves. Celery salt is used as a seasoning, in cocktails (notably to enhance the flavour of Bloody Mary cocktails), on the Chicago-style hot dog, and in Old Bay Seasoning.

Herbalism

The use of celery seed in pills for relieving pain was described by Aulus Cornelius Celsus around AD 30. Celery seeds contain a compound, 3-*n*-butylphthalide, that has been demonstrated to lower blood pressure in rats.

Celery seeds

Celery juice significantly reduced hypertension in 87.5% of patients (14 of 16) tested. Another study showed the same effect on hypertension associated with pregnancy.

There is evidence that suggests a link between celery extract and male reproductive health, spermatogenesis, and testicular damage reversal, but this link has yet to be tested on human subjects.

Bergapten in the seeds can increase photosensitivity, so the use of essential oil externally in bright sunshine should be avoided. The oil and large doses of seeds should be avoided during pregnancy, as they can act as a uterine stimulant. Seeds intended for cultivation are not suitable for eating as they are often treated with fungicides.

Nutrition

Celery is used in weight-loss diets, where it provides low-calorie dietary fibre bulk. Celery is often incorrectly thought to be a "negative-calorie food," the digestion of which burns more calories than the body can obtain. In fact, eating celery provides positive net calories, with digestion consuming only a small proportion of the calories taken in.

Allergies

Celery is among a small group of foods (headed by peanuts) that appear to provoke the most severe allergic reactions; for people with celery allergy, exposure can cause potentially fatal anaphylactic shock. The allergen does not appear to be destroyed at cooking temperatures. Celery root—commonly eaten as celeriac, or put into drinks—is known to contain more allergen than the stalk. Seeds contain the highest levels of allergen content. Exercise-induced anaphylaxis may be exacerbated. An allergic reaction also may be triggered by eating foods that have been processed with machines that have previously processed celery, making avoiding such foods difficult. In contrast with peanut allergy being most prevalent in the US, celery allergy is most prevalent in Central Europe. In the European Union, foods that contain or may contain celery, even in trace amounts, must be clearly marked as such.

Chemistry

Polyynes can be found in Apiaceae vegetables like celery, and their extracts show cytotoxic activities. Celery contains phenolic acid, which is an antioxidant.

Apiin and apigenin can be extracted from celery and parsley. Lunularin is a dihydrostilbenoid found in common celery.

Some aromatic compounds of celery leaves and stalks are reported as butylphthalide and sedanolide.

History

Selinunte didrachm coin bearing a *selinon* (celery) leaf, circa 515-470 BC.

Daniel Zohary and Maria Hopf note that celery leaves and inflorescences were part of the garlands found in the tomb of pharaoh Tutankhamun (died 1323 BC), and celery mericarps dated to the seventh century BC were recovered in the Heraion of Samos. However, they note "since *A. graveolens* grows wild in these areas, it is hard to decide whether these remains represent wild or cultivated forms." Only by classical times is it certain that celery was cultivated.

M. Fragiska mentions an archeological find of celery dating to the 9th century BC, at Kastanas; however, the literary evidence for ancient Greece is far more abundant. In Homer's *Iliad*, the horses of the Myrmidons graze on wild celery that grows in the marshes of Troy, and in *Odyssey*, there is mention of the meadows of violet and wild celery surrounding the cave of Calypso.

In the *Capitulary* of Charlemagne, compiled ca. 800, *apium* appears, as does *olisatum*, or alexanders, among medicinal herbs and vegetables the Frankish emperor desired to see grown. At some later point in medieval Europe celery displaced alexanders.

Celery's late arrival in the English kitchen is an end-product of the long tradition of seed selection needed to reduce the sap's bitterness and increase its sugars. By 1699, John Evelyn could recommend it in his *Acetaria. A Discourse of Sallets*: "Sellery, apium Italicum, (and of the Petroseline Family) was formerly a stranger with us (nor very long since in Italy) is an hot and more generous sort of Macedonian Persley or Smallage...and for its high and grateful Taste is ever plac'd in the middle of the Grand Sallet, at our Great Men's tables, and Praetors feasts, as the Grace of the whole Board".

Celery makes a minor appearance in colonial American gardens; its culinary limitations are reflected in the observation by the author of *A Treatise on Gardening, by a Citizen of Virginia* that it is "one of the species of parsley." Its first extended treatment in print was in Bernard M'Mahon's *American Gardener's Calendar* (1806). After the mid-19th century, continued selections for refined crisp texture and taste brought celery to American tables, where it was served in celery vases to be salted and eaten raw.

Cultural Depictions

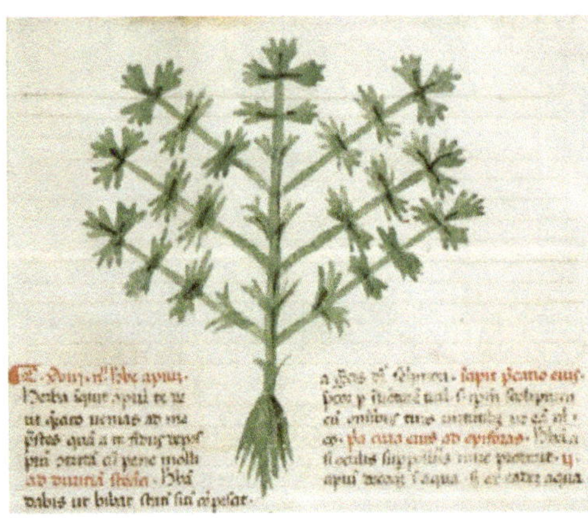

Apium illustration from Barbarus Apuleius' *Herbarium*, c. 1400.

A chthonian symbol among the ancient Greeks, celery was said to have sprouted from the blood of Kadmilos, father of the Cabeiri, chthonian divinities celebrated in Samothrace, Lemnos, and

Thebes. The spicy odour and dark leaf colour encouraged this association with the cult of death. In classical Greece, celery leaves were used as garlands for the dead, and the wreaths of the winners at the Isthmian Games were first made of celery before being replaced by crowns made of pine. According to Pliny the Elder in Achaea, the garland worn by the winners of the sacred Nemean Games was also made of celery. The Ancient Greek colony of Selinous (Greek *Selinous*), on Sicily, was named after wild parsley that grew abundantly there; Selinountian coins depicted a parsley leaf as the symbol of the city.

The perennial BBC television series Doctor Who featured the Fifth Doctor (played by Peter Davison, from 1981–84), who wore a sprig of celery as a corsage.

The name "celery" retraces the plant's route of successive adoption in European cooking, as the English "celery" (1664) is derived from the French *céleri* coming from the Lombard term, *seleri*, from the Latin *selinon*, borrowed from Greek.

References

- Durie, Sir Mason (2010). "Te whakahaumanutanga me te oranga hinengaro o mua – Ka mātaia ngā huanga o te rongoā – Traditional healing and mental health: measuring the effectiveness of rongoā" (PDF). Best Practice Journal. Best Practice Advocacy Centre. June (28): 5–7
- Takele, Etaferahu. "Celery Production: Sample Costs and Profitability Analysis" (PDF). UC Davis. University of California Agriculture and Natural Resources. Retrieved 16 April 2017
- Jones, Daniel (2003) [1917], Peter Roach; James Hartmann; Jane Setter, eds., English Pronouncing Dictionary, Cambridge: Cambridge University Press, ISBN 3-12-539683-2
- Srivastava, JK; Shankar, E; Gupta, S (November 2010). "Chamomile: A herbal medicine of the past with bright future". Mol Med Rep. 3: 895–901. PMC 2995283. PMID 21132119. doi:10.3892/mmr.2010.377
- "Sophora microphylla (Kowhai)". Taranaki Educational Resource: Research, Analysis and Information Network. Retrieved 31 August 2012
- Abebe, W. (2002-12-01). "Herbal medication: potential for adverse interactions with analgesic drugs". Journal of Clinical Pharmacy and Therapeutics. 27 (6): 391–401. ISSN 0269-4727. PMID 12472978
- Conti, Fabio; Carlo Blasi; Alessandro Alessandrini; Giovanna Abbate (May 2005). An annotated Checklist of the Italian Vascular Flora. p. 60. ISBN 88-7621-458-5
- Jones, Rhys (2 March 2009). "Rongoā – medicinal use of plants – Other medicinal plants". Te Ara – the Encyclopedia of New Zealand. Retrieved 8 November 2010
- Baumann, LS (2007). "Less-known botanical cosmeceuticals". Dermatologic therapy. 20 (5): 330–342. PMID 18045358. doi:10.1111/j.1529-8019.2007.00147.x
- Rizzo, V (January 2009). "Effects of packaging on shelf life of fresh celery". Journal of Food Engineering. 90 (1): 124–128. Retrieved 21 April 2017
- Newman, Jacqueline (Fall 2006). "Chinese Celery". Vegetables and Vegetarian Foods. 13 (3): 15–34. Retrieved 17 April 2017
- Nestle, M.; Nesheim, M.C. (2012). Why Calories Count: From Science to Politics. University of California Press. p. 189. ISBN 9780520262881. Retrieved 2014-10-05
- "BSBI List 2007". Botanical Society of Britain and Ireland. Archived from the original (xls) on 25 January 2015. Retrieved 2014-10-17
- Minto, Robert E.; Blacklock, Brenda J "Biosynthesis and function of polyacetylenes and allied natural products" From Progress in Lipid Research 2008, vol. 47, 233-306. doi:10.1016/j.plipres.2008.02.002

3

Medicinal Plants used in Tradional Medicine

Medicinal plants have been used to treat patients in various traditions like traditional African medicine, traditional Chinese medicine and Ayurveda. Traditional Chinese medicine includes herbs such as Angelica sinensis, goji, Astragalus, Lingzhi mushroom, etc. The topics discussed in the chapter are of great importance to broaden the existing knowledge on medicinal plants.

Ayurveda

Ayurveda or Ayurveda medicine, is a system of medicine with historical roots in the Indian subcontinent. Globalized and modernized practices derived from Ayurveda traditions are a type of complementary or alternative medicine. In the Western world, Ayurveda therapies and practices (which are manifold) have been integrated in general wellness applications and as well in some cases in medical use.

The main classical Ayurveda texts begin with accounts of the transmission of medical knowledge from the Gods to sages, and then to human physicians. In *Sushruta Samhita* ([*Sushruta's Compendium*]), Sushruta wrote that Dhanvantari, Hindu god of Ayurveda, incarnated himself as a king of Varanasi and taught medicine to a group of physicians, including Sushruta. Ayurveda therapies have varied and evolved over more than two millennia. Therapies are typically based on complex herbal compounds, minerals and metal substances (perhaps under the influence of early Indian alchemy or *rasa shastra*). Ancient Ayurveda texts also taught surgical techniques, including rhinoplasty, kidney stone extractions, sutures, and the extraction of foreign objects.

Although laboratory experiments suggest it is possible that some substances used in Ayurveda might be developed into effective treatments, there is no evidence that any are effective as currently practiced. Ayurveda medicine is considered pseudoscientific. Other researchers consider it a protoscience, or trans-science system instead. Close to 21% of Ayurveda U.S. and Indian-manufactured patent medicines sold through the Internet were found to contain toxic levels of heavy metals, specifically lead, mercury, and arsenic. The public health implications of such metallic contaminants in India are unknown.

Some scholars assert that Ayurveda originated in prehistoric times, and that some of the concepts of Ayurveda have existed from the time of the Indus Valley Civilization or even earlier. Ayurveda developed significantly during the Vedic period and later some of the non-Vedic systems such as Buddhism and Jainism also developed medical concepts and practices that appear in the classical Ayurveda texts. Humoral balance is emphasized, and suppressing natural urges is considered unhealthy and claimed to lead to illness. Ayurveda names three elemental substances, the *doshas* (called Vata, Pitta and Kapha), and states that a balance of the doshas results in health, while imbalance results in disease. Ayurveda has eight canonical components, which are derived from classical Sanskrit literature. Ayurveda practitioners had developed various medicinal preparations and surgical procedures by the medieval period.

Eight Components

The earliest classical Sanskrit works on Ayurveda describe medical science as being divided into eight components (Skt. *aṅga*). This characterization of the physicians' art, "the medicine that has eight components" is first found in the Sanskrit epic the *Mahābhārata*, ca 4th cenury BCE. The components are:

- *Kāyacikitsā*: general medicine, medicine of the body
- *Kaumāra-bhṛtya*: the treatment of children, paediatrics
- *Śalyatantra*: surgical techniques and the extraction of foreign objects
- Śālākyatantra: treatment of ailments affecting ears, eyes, nose, mouth, etc. ("ENT")
- *Bhūtavidyā*: pacification of possessing spirits, and the people whose minds are affected by such possession
- *Agadatantra*: toxicology
- *Rasāyanatantra*: rejuvenation and tonics for increasing lifespan, intellect and strength
- *Vājīkaraṇatantra*: aphrodisiacs and treatments for increasing the volume and viability of semen and sexual pleasure.

Principles and Terminology

Shown in the image is the philosopher Nagarjuna, known chiefly for his doctrine of the *Madhyamaka* (middle path). He wrote the medical works *The Hundred Prescriptions* and *The Precious Collection*, among others.

The central theoretical ideas of Ayurveda developed in the mid-first millennium BCE, and show parallels with Sāṅkhya and Vaiśeṣika philosophies, as well as with Buddhism and Jainism. Balance is emphasized, and suppressing natural urges is considered unhealthy and claimed to lead to illness. For example, to suppress sneezing is said to potentially give rise to shoulder pain. However, people are also cautioned to stay within the limits of reasonable balance and measure when following nature's urges. For example, emphasis is placed on moderation of food intake, sleep, and sexual intercourse.

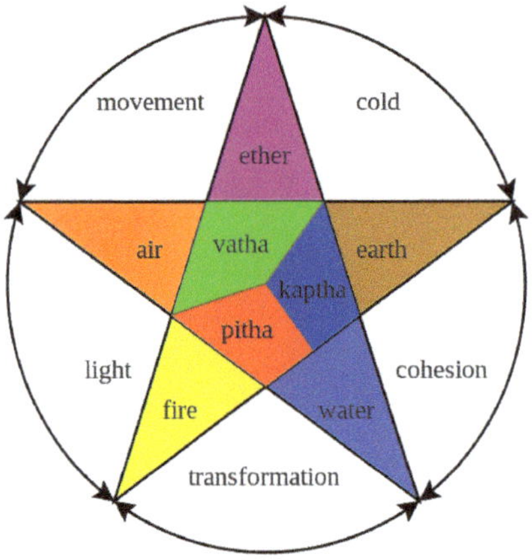

The three doṣhas and the five elements from which they are composed.

Ayurveda names seven basic tissues (dhatu), which are plasma (*rasa*), blood (*rakta*), muscles (*māmsa*), fat (*meda*), bone (*asthi*), marrow (*majja*), and semen (*shukra*). Like the medicine of classical antiquity, Ayurveda has historically divided bodily substances into five classical elements (Sanskrit *[maha]panchabhuta*, viz. earth, water, fire, air and ether. There are also twenty gunas (qualities or characteristics) which are considered to be inherent in all substances. These are organized in ten pairs: heavy/light, cold/hot, unctuous/dry, dull/sharp, stable/mobile, soft/hard, non-slimy/slimy, smooth/coarse, minute/gross, and viscous/liquid.

Ayurveda also names three elemental substances, the *doshas* (called Vata, Pitta and Kapha), and states that a balance of the doshas results in health, while imbalance results in disease. One Ayurvedic view is that the doshas are balanced when they are equal to each other, while another view is that each human possesses a unique combination of the doshas which define this person's temperament and characteristics. In either case, it says that each person should modulate their behavior or environment to increase or decrease the doshas and maintain their natural state.

In medieval taxonomies of the Sanskrit knowledge systems, Ayurveda is assigned a place as a subsidiary Veda (upaveda). Some medicinal plant names from the *Atharvaveda* and other Vedas can be found in subsequent Ayurveda literature. The earliest recorded theoretical statements about the canonical models of disease in Ayurveda occur in the earliest Buddhist Canon.

Practice

Ayurvedic doctors regard physical existence, mental existence, and personality as a unit, with each element being able to influence the others. This is a holistic approach used during diagnosis and therapy, and is a fundamental aspect of Ayurveda. Another part of Ayurvedic treatment says that there are channels (*srotas*) which transport fluids, and that the channels can be opened up by massage treatment using oils and Swedana (fomentation). Unhealthy channels are thought to cause disease.

Diagnosis

Ayurveda has eight ways to diagnose illness, called Nadi (pulse), Mootra (urine), Mala (stool), Jihva (tongue), Shabda (speech), Sparsha (touch), Druk (vision), and Aakruti (appearance). Ayurvedic practitioners approach diagnosis by using the five senses. For example, hearing is used to observe the condition of breathing and speech. The study of the lethal points or *marman marma* is of special importance.

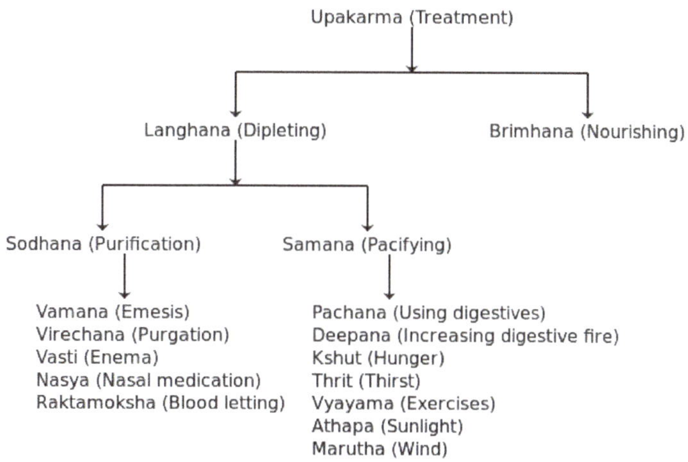

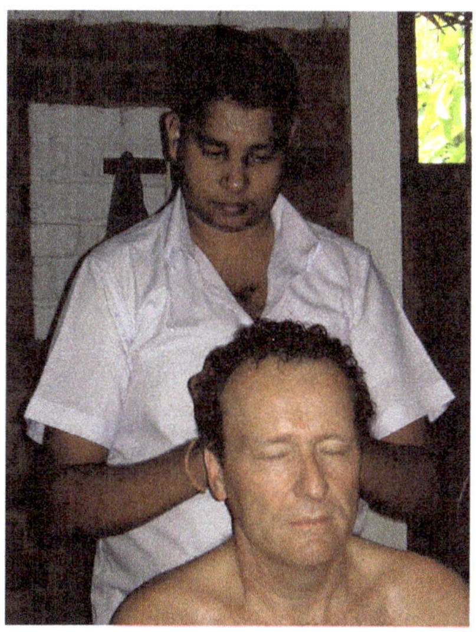

An Ayurvedic practitioner applying oils using head massage.

Treatment and Prevention

Two of the eight branches of classical Ayurveda deal with surgery (*Śalya-cikitsā* and *Śālākya-tantra*), but contemporary Ayurveda tends to stress attaining vitality by building a healthy metabolic system and maintaining good digestion and excretion. Ayurveda also focuses on exercise, yoga, and meditation. One type of prescription is a Sattvic diet.

Ayurveda follows the concept of Dinacharya, which says that natural cycles (waking, sleeping, working, meditation etc.) are important for health. Hygiene, including regular bathing, cleaning of teeth, skin care, and eye washing, is also a central practice.

Substances Used

Plant-based treatments in Ayurveda may be derived from roots, leaves, fruits, bark, or seeds such as cardamom and cinnamon. In the 19th century, William Dymock and co-authors summarized hundreds of plant-derived medicines along with the uses, microscopic structure, chemical composition, toxicology, prevalent myths and stories, and relation to commerce in British India. Animal products used in Ayurveda include milk, bones, and gallstones. In addition, fats are prescribed both for consumption and for external use. Consumption of minerals, including sulphur, arsenic, lead, copper sulfate and gold, are also prescribed. The addition of minerals to herbal medicine is called *rasa shastra*.

Ayurveda uses alcoholic beverages called *Madya*, which are said to adjust the doshas by increasing Pitta and reducing Vatta and Kapha. Madya are classified by the raw material and fermentation process, and the categories include: sugar-based, fruit-based, cereal-based, cereal-based with herbs, fermentated with vinegar, and tonic wines. The intended outcomes can include causing purgation, improving digestion or taste, creating dryness, or loosening joints. Ayurvedic texts describe Madya as non-viscid and fast-acting, and say that it enters and cleans minute pores in the body.

Purified opium is used in eight Ayurvedic preparations and is said to balance the Vata and Kapha doshas and increase the Pitta dosha. It is prescribed for diarrhea and dysentery, for increasing the sexual and muscular ability, and for affecting the brain. The sedative and pain-relieving properties of opium are not considered in Ayurveda. The use of opium is not found in the ancient Ayurvedic texts, and is first mentioned in the *Sarngadhara Samhita* (1300-1400 CE), a book on pharmacy used in Rajasthan in Western India, as an ingredient of an aphrodisiac to delay male ejaculation. It is possible that opium was brought to India along with or before the Mohammedan conquest. The book *Yoga Ratnakara* (1700-1800 CE, unknown author), which is popular in Maharashtra, uses opium in a herbal-mineral composition prescribed for diarrhea. In the *Bhaisajya Ratnavali*, opium and camphor are used for acute gastroenteritis. In this drug, the respiratory depressant action of opium is counteracted by the respiratory stimulant property of Camphor. Later books have included the narcotic property for use as analgesic pain reliever.

Cannabis indica is also absent from the ancient Ayurveda books, and is first mentioned in the *Sarngadhara Samhita* as a treatment for diarrhea. In the *Bhaisajya Ratnavali* it is named as an ingredient in an aphrodisiac.

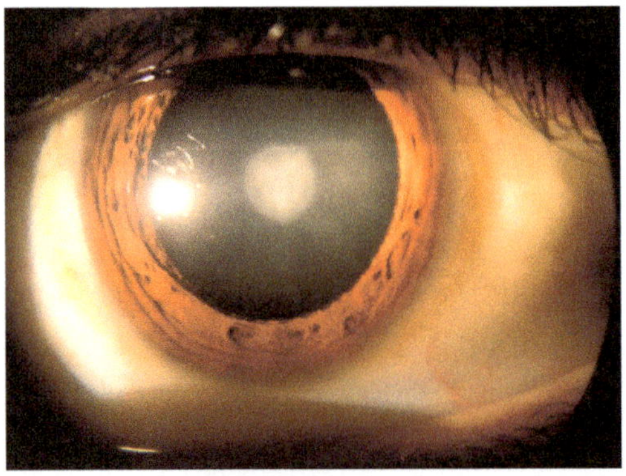

Cataract in human eye – magnified view seen on examination with a slit lamp. Cataract surgery is mentioned in the *Sushruta Samhita*, as a procedure to be performed with a *jabamukhi salaka*, a curved needle used to loosen the obstructing phlegm and push it out of the field of vision. The eye would later be soaked with warm butter and then bandaged.

Ayurveda says that both oil and tar can be used to stop bleeding, and that traumatic bleeding can be stopped by four different methods: ligation of the blood vessel, cauterisation by heat, use of preparations to facilitate clotting, and use of preparations to constrict the blood vessels. Oils are also used in a number of ways, including regular consumption, anointing, smearing, head massage, application to affected areas, and oil pulling. Liquids may also be poured on the patient's forehead, a technique called shirodhara.

Panchakarma

According to Ayurveda, the technique of panchakarma (Devanāgarī: पंचकर्म)) eliminates toxic elements from the body. Panchakarma includes Vamana, Virechana, Basti, Nasya and Raktamokshana. It is preceded by Poorvakarma as a preparatory step, and is followed by Paschatkarma and Peyadikarma.

Current Status

A typical Ayurvedic pharmacy, Rishikesh.

Based on the World Health Assembly resolution on traditional medicine (WHA62.13, compare as well the 2008 Beijing declaration on traditional medicine), the WHO has established a traditional medicine strategy, which involves, among other complementary health systems, Ayurveda.The first previous global strategy was published in 2002 and dealt especially with herbal medicines. The current version (2014-2023) is beginning to consider aspects of T&CM practices and practitioners and whether and how they should be integrated into overall health service delivery. It is as well part of the 2013 WHO, WIPO, WTO TRIPS study on health innovation and access to medicines.

The WTO mentions Ayurveda in its intellectual property strategy. Traditional medicine, including Ayurveda, contributes significantly to the health status of many communities, and is increasingly used within certain communities in developed countries. Traditional medicine has a long history of use in health maintenance and in disease prevention and treatment, particularly for chronic diseases. Suitable recognition of traditional medicine is an important element of national health policies and has as well important basis for new products with significant export potential. WTO, WIPO and WHO ask for respect for both the economic value and the social and cultural significance of traditional knowledge and supports documentation of traditional medical knowledge and enhanced regulation of quality, safety and efficacy of such products. The WHO paper asks to integrate traditional and complementary medicine services, where appropriate, into health care service delivery and self-health care and to promote an improved universal health coverage by doing so.

India

According to some sources, up to 80 percent of people in India use some form of traditional medicine, a category which includes Ayurveda.

In 1970, the Indian Medical Central Council Act which aimed to standardise qualifications for Ayurveda practitioners and provide accredited institutions for its study and research was passed by the Parliament of India. In 1971, the Central Council of Indian Medicine (CCIM) was established under the Department of Ayurveda, Yoga and Naturopathy, Unani, Siddha and Homoeopathy (AYUSH), Ministry of Health and Family Welfare, to monitor higher education in Ayurveda in India. The Indian government supports research and teaching in Ayurveda through many channels at both the national and state levels, and helps institutionalise traditional medicine so that it can be studied in major towns and cities. The state-sponsored Central Council for Research in Ayurvedic Sciences (CCRAS) is designed to do research on Ayurveda. Many clinics in urban and rural areas are run by professionals who qualify from these institutes. As of 2013, India has over 180 training centers offer degrees in traditional Ayurvedic medicine.

To fight biopiracy and unethical patents, in 2001 the government of India set up the Traditional Knowledge Digital Library as a repository for formulations of various systems of Indian medicine, such as Ayurveda, Unani and Siddha. The formulations come from over 100 traditional Ayurveda books. An Indian Academy of Sciences document quoting a 2003-04 report states that India had 432,625 registered medical practitioners, 13,925 dispensaries, 2,253 hospitals and a bed strength of 43,803. 209 under-graduate teaching institutions and 16 post-graduate institutions. Insurance companies cover expenses for Ayurvedic treatments in case of conditions such as spinal cord disorders, bone disorder, arthritis and cancer. Such claims comprise 5-10 percent of the country's health insurance claims.

Maharashtra Andhashraddha Nirmoolan Samiti, an organisation dedicated to fighting superstition in India, considers Ayurveda to be pseudoscience.

Other Countries on the Indian Subcontinent

Tulsi-flower (holy basil), an Ayurvedic herb

About 75%-80% of the population of Nepal use Ayurveda, and it is the most practiced form of medicine in the country.

The Sri Lankan tradition of Ayurveda is similar to the Indian tradition. Practitioners of Ayurveda in Sri Lanka refer to Sanskrit texts which are common to both countries. However, they do differ in some aspects, particularly in the herbs used.

In 1980, the Sri Lankan government established a Ministry of Indigenous Medicine to revive and regulate Ayurveda. The Institute of Indigenous Medicine (affiliated to the University of Colombo) offers undergraduate, postgraduate, and MD degrees in Ayurveda Medicine and Surgery, and similar degrees in unani medicine. In the public system, there are currently 62 Ayurvedic hospitals and 208 central dispensaries, which served about 3 million people (about 11% of Sri Lanka's population) in 2010. In total, there are about 20,000 registered practitioners of Ayurveda in the country.

According to the Mahavamsa, an ancient chronicle of Sinhalese royalty from the sixth century C.E., King Pandukabhaya of Sri Lanka (reigned 437 BCE to 367 BCE) had lying-in-homes and Ayurvedic hospitals (Sivikasotthi-Sala) built in various parts of the country. This is the earliest documented evidence available of institutions dedicated specifically to the care of the sick anywhere in the world. Mihintale Hospital is the oldest in the world.

Outside the Indian Subcontinent

Ayurveda is a system of traditional medicine developed during antiquity and the medieval period, and as such is comparable to pre-modern Chinese and European systems of medicine. However, beginning in the 1960s, Ayurveda has been advertised as alternative medicine in the Western world. Due to different laws and medical regulations in the rest of the world, the unregulated practice and commercialisation of Ayurveda has raised ethical and legal issues. In some instances, Ayurvedic practices or terminology have also been adapted specifically for Western consumption, notably in the case of "Maharishi Ayurveda" in the 1980s. In some cases, this involved active fraud on the part of proponents of Ayurveda in an attempt to falsely represent the system as equal to the standards of modern medical research.

Baba Hari Dass was an early proponent who helped bring Ayurveda to the US in the early 1970s. He taught classes derived from the *Suśrutha Saṃhitā* and the *Charaka Saṃhitha*, leading to the establishment of the Mount Madonna Institute, College of Ayurveda, Ayurveda World, and Ayurvedic pharmacy. He invited several notable Ayurvedic teachers, including Vasant Lad, Sarita Shrestha, and Ram Harsh Singh. The Ayurvedic practitioner Michael Tierra wrote that "the history of Ayurveda in North America will always owe a debt to the selfless contributions of Baba Hari Dass."

In the United States, the practice of Ayurveda is not licensed or regulated by any state. Practitioners of Ayurveda can be licensed in other healthcare fields such as massage therapy or midwifery, and a few states have approved schools teaching Ayurveda.

Classification and Efficacy

Although laboratory experiments suggest it is possible that some substances in Ayurveda might be developed into effective treatments, there is no evidence that any are effective in themselves. According to Cancer Research UK, no significant scientific evidence has shown effectiveness of

Ayurvedic medicine for the treatment of any disease, although massage and relaxation are often beneficial for some cancer patients and there are indications from animal studies that some herbal products used in Ayurveda might be explored further.

Today, ayurvedic medicine is considered pseudoscientific on account of its confusion between reality and metaphysical concepts. Other researchers debate whether it should be considered a protoscience, an unscientific, or trans-science system instead.

A review of the use of Ayurveda for cardiovascular disease concluded that the evidence is not convincing for the use of any Ayurvedic herbal treatment for heart disease or hypertension, but that many herbs used by Ayurvedic practitioners could be appropriate for further research.

Research into ayurveda has been characterized as pseudoscience. Both the lack of scientific soundness in the theoretical foundations of ayurveda and the quality of research have been criticized.

Research

In India, research in Ayurveda is undertaken by the Ministry of AYUSH, an abbreviation for the Department of Ayurveda, Yoga and Naturopathy, Unani, Siddha and Homoeopathy, through a national network of research institutes.

In Nepal, the National Ayurvedic Training and Research Centre (NATRC) researches medicinal herbs in the country.

In Sri Lanka, the Ministry of Health, Nutrition and Indigenous Medicine looks after the research in Ayurveda through various national research institutes.

Use of Toxic Metals

Rasa shastra, the practice of adding metals, minerals or gems to herbal preparations, may include toxic heavy metals such as lead, mercury and arsenic. The public health implications of metals in rasa shastra in India is unknown. Adverse reactions to herbs are described in traditional Ayurvedic texts, but practitioners are reluctant to admit that herbs could be toxic and that reliable information on herbal toxicity is not readily available. There is a communication gap between practitioners of modern medicine and Ayurveda.

Some traditional Indian herbal medicinal products contain harmful levels of heavy metals, including lead. A 1990 study on Ayurvedic medicines in India found that 41% of the products tested contained arsenic, and that 64% contained lead and mercury. A 2004 study found toxic levels of heavy metals in 20% of Ayurvedic preparations made in South Asia and sold in the Boston area, and concluded that Ayurvedic products posed serious health risks and should be tested for heavy-metal contamination. A 2008 study of more than 230 products found that approximately 20% of remedies (and 40% of *rasa shastra* medicines) purchased over the Internet from U.S. and Indian suppliers contained lead, mercury or arsenic.

Heavy metals are thought of as active ingredients by advocates of Indian herbal medicinal products. According to ancient Ayurvedic texts, certain physico-chemical purification processes such as *samskaras* or *shodhanas* (for metals) 'detoxicify' the heavy metals in it. These are similar to the Chinese *pao zhi*, although the Ayurvedic techniques are more complex and may involve physical pharmacy techniques as well as mantras. However, these products have nonetheless caused severe lead poisoning and other toxic effects. Between 1978 and 2008, "more than 80 cases of lead poisoning associated with Ayurvedic medicine use [were] reported worldwide". In 2012, the U.S. Centers for Disease Control and Prevention (CDC) linked Ayurvedic drugs to lead poisoning, based on cases where toxic materials were found in the blood of pregnant women who had taken Ayurvedic drugs.

Ayurvedic practitioners argue that the toxicity of bhasmas comes from improper manufacturing processes, contaminants, improper use of Ayurvedic medicine, quality of raw materials and that the end products and improper procedures are used by charlatans.

In India, the government ruled that Ayurvedic products must be labelled with their metallic content. However, in *Current Science*, a publication of the Indian Academy of Sciences, M. S. Valiathan said that "the absence of post-market surveillance and the paucity of test laboratory facilities [in India] make the quality control of Ayurvedic medicines exceedingly difficult at this time". In the United States, most Ayurvedic products are marketed without having been reviewed or approved by the FDA. Since 2007, the FDA has placed an import alert on some Ayurvedic products in order to prevent them from entering the United States.

History

The origins of Ayurveda have been traced back to around 5,000 BCE, when they originated as an oral tradition. Some of the concepts of Ayurveda have existed since the times of Indus Valley Civilization. The first recorded forms of Ayurveda as medical texts evolved from the Vedas. Ayurveda is a discipline of the upaveda or "auxiliary knowledge" in Vedic tradition. The origins of Ayurveda are also found in Atharvaveda, which contains 114 hymns and incantations described as magical cures for disease. There are various legendary accounts of the origin of Ayurveda, e.g. that it was received by Dhanvantari (or Divodasa) from Brahma. Tradition also holds that the writings of Ayurveda were influenced by a lost text by the sage Agnivesa.

Charaka

Ayurveda is one of the few systems of medicine developed in ancient times that is still widely practiced in modern times. As such, it is open to the criticism that its conceptual basis is obsolete and that its contemporary practitioners have not taken account of the developments of modern medicine. Responses to this situation led to an impassioned debate in India during the early decades of the twentieth century, between proponents of unchanging tradition (*śuddha* "pure" ayurveda) and those who thought ayurveda should modernise and syncretize (*aśuddha* "impure, tainted" ayurveda). The political debate about the place of ayurveda in contemporary India has continued to the present (2015), both in the public arena and in government. Debate about the place of Ayurvedic medicine in the contemporary internationalized world also continues today (2015).

Main Texts

There are three principal early texts on Ayurveda, the Charaka Samhita, the Sushruta Samhita and the Bhela Samhita. The Sushruta Samhita is based on an original from the 6th century BCE, and was updated by the Buddhist scholar Nagarjuna in the 2nd century CE. The Charaka Samhita, written by Charaka, and the Bhela Samhita, attributed to Atreya Punarvasu, are also dated to the 6th century BCE. The Charaka Samhita was also updated by Dridhabala during the early centuries of the Common Era.

The Bower Manuscript includes of excerpts from the Bheda Samhita and its description of concepts in Central Asian Buddhism. In 1987, A. F. R. Hoernle identified the scribe of the medical portions of the manuscript to be a native of India using a northern variant of the Gupta script, who had migrated and become a Buddhist monk in a monastery in Kucha. The Chinese pilgrim Fa Hsien (c. 337–422 AD) wrote about the healthcare system of the Gupta empire (320–550) and described the institutional approach of Indian medicine. This is also visible in the works of Charaka, who describes hospitals and how they should be equipped.

Other early texts are the Agnivesha Samhita, Kasyapa Samhita and Harita Samhita. The original edition of the Agnivesha Samhita, by Agnivesa, is dated to 1500 BCE, and it was later modified by Charaka. Kasyapa Samhita includes the treatise of Jivaka Kumar Bhaccha and is dated to the 6th century BCE. While Harita Samhita is dated to an earlier period, it is attributed to Harita, who was a disciple of Punarvasu Atreya. Some later texts are *Astanga nighantu* (8th Century) by Vagbhata, *Paryaya ratnamala* (9th century) by Madhava, *Siddhasara nighantu* (9th century) by Ravi Gupta, *Dravyavali* (10th Century), and *Dravyaguna sangraha* (11th century) by Cakrapanidatta, among others.

Illnesses Portrayed

Underwood and Rhodes state that the early forms of traditional Indian medicine identified fever, cough, consumption, diarrhea, dropsy, abscesses, seizures, tumours, and leprosy, and that treatments included plastic surgery, lithotomy, tonsillectomy, couching (a form of cataract surgery), puncturing to release fluids in the abdomen, extraction of foreign bodies, treatment of anal fistulas, treating fractures, amputations, cesarean sections, and stitching of wounds. The use of herbs and surgical instruments became widespread. During this period, treatments were also prescribed for complex ailments, including angina pectoris, diabetes, hypertension, and stones.

Further Development and Spread

Ayurveda flourished throughout the Indian Middle Ages. Dalhana (fl. 1200), Sarngadhara (fl. 1300) and Bhavamisra (fl. 1500) compiled works on Indian medicine. The medical works of both Sushruta and Charaka were also translated into the Chinese language in the 5th century, and during the 8th century, they were translated into the Arabic and Persian language. The 9th-century Persian physician Muhammad ibn Zakariya al-Razi was familiar with the text. The Arabic works derived from the Ayurvedic texts eventually also reached Europe by the 12th century. In Renaissance Italy, the Branca family of Sicily and Gaspare Tagliacozzi (Bologna) were influenced by the Arabic reception of the Sushruta's surgical techniques.

British physicians traveled to India to observe rhinoplasty being performed using native methods, and reports on Indian rhinoplasty were published in the *Gentleman's Magazine* in 1794. Instruments described in the *Sushruta Samhita* were further modified in Europe. Joseph Constantine Carpue studied plastic surgery methods in India for 20 years and, in 1815, was able to perform the first major rhinoplasty surgery in the western world, using the "Indian" method of nose reconstruction. In 1840 Brett published an article about this technique.

During the period of colonial British rule of India, the practice of Ayurveda was neglected by the British Indian Government, in favor of modern medicine. After Indian independence, there was more focus on Ayurveda and other traditional medical systems. Ayurveda became a part of the Indian National health care system, with state hospitals for Ayurveda established across the country. However, the treatments of traditional medicines were not always integrated with others.

Ocimum Tenuiflorum

Ocimum tenuiflorum (synonym *Ocimum sanctum*), commonly known as holy basil, *tulasi* (sometimes spelled *thulasi*) or *tulsi*, is an aromatic plant in the family Lamiaceae which is native to the Indian subcontinent and widespread as a cultivated plant throughout the Southeast Asian tropics.

Tulasi is cultivated for religious and medicinal purposes, and for its essential oil. It is widely known across the Indian subcontinent as a medicinal plant and a herbal tea, commonly used in Ayurveda, and has an important role within the Vaishnava tradition of Hinduism, in which devotees perform worship involving holy basil plants or leaves. This plant is revered as an elixir of life.

The variety of *Ocimum tenuiflorum* used in Thai cuisine is referred to as Thai holy basil (*kaphrao*); it is different than Thai basil, which is a variety of *Ocimum basilicum*.

Morphology

Close-up of *tulasi* leaves

Holy basil is an erect, many-branched subshrub, 30–60 cm (12–24 in) tall with hairy stems. Leaves are green or purple; they are simple, petioled, with an ovate, up to 5 cm (2.0 in)-long blade which usually has a slightly toothed margin; they are strongly scented and have a decussate phyllotaxy. The purplish flowers are placed in close whorls on elongate racemes. The two main morphotypes cultivated in India and Nepal are green-leaved (Sri or Lakshmi *tulasi*) and purple-leaved (Krishna *tulasi*).

Tulasi flowers

Origin and Distribution

DNA barcodes of various biogeographical isolates of tulsi from the Indian subcontinent are now available. In a large-scale phylogeographical study of this species conducted using chloroplast genome sequences, a group of researchers from Central University of Punjab, Bathinda, have found that this plant originates from North Central India. The discovery might suggest the evolution of tulsi is related with the cultural migratory patterns in the Indian subcontinent.

Uses

In Hinduism

Tulsi leaves are an essential part in the worship of Vishnu and his avatars, including Krishna and Ram, and other male Vaishnava deities such as Hanuman, Balarama, Garuda and many others. *Tulsi* is a sacred plant for Hindus and is worshipped as the avatar of Lakshmi. It is believed that water mixed with the petals given to the dying raises their departing souls to heaven. *Tulsi*, which is Sanskrit for "the incomparable one", is most often regarded as a consort of Krishna in the form of Lakshmi. According to the *Brahma Vaivarta Purana*, *tulsi* is an expression of Sita. There are two types of tulsi worshipped in Hinduism: "Rama tulsi" has light green leaves and is larger in size; "Shyama tulsi" has dark green leaves and is important for the worship of Hanuman. Many Hindus have tulasi plants growing in front of or near their home, often in special pots. Traditionally, *tulsi* is planted in the centre of the central courtyard of Hindu houses. It is also frequently grown next to Hanuman temples, especially in Varanasi.

According to Vaishnavas, it is believed in Puranas that during Samudra Manthana, when the gods win the ocean-churning against the asuras, Dhanvantari comes up from the ocean with Amrit in hand for the gods. Dhanvantari, the divine healer, sheds happy tears, and when the first drop falls in the Amrit, it forms tulasi. In the ceremony of Tulsi Vivaha, *tulsi* is ceremonially married to Krishna annually on the eleventh day of the waxing moon or twelfth of the month of *Kartik* in the lunar calendar. This day also marks the end of the four-month Chaturmas, which is considered inauspicious for weddings and other rituals, so the day inaugurates the annual marriage season in India. The ritual lighting of lamps each evening during *Kartik* includes the worship of the *tulsi* plant, which is held to be auspicious for the home. Vaishnavas especially follow the daily worship of tulsi during Kartik. In another legend, Tulsi was a pious woman who sought a boon to marry Vishnu. Lakshmi, Vishnu's consort, cursed her to become a plant in earth. However, Vishnu appeased her by giving her a boon that she would grace him when he appears in the form of Shaligrama in temples.

Vaishnavas traditionally use Hindu prayer beads made from tulsi stems or roots, which are an important symbol of initiation. Tulsi rosaries are considered to be auspicious for the wearer, and believed to put them under the protection of Hanuman. They have such a strong association with Vaishnavas, that followers of Hanuman are known as "those who bear the tulsi round the neck".

Tulsi grown in front of a house

Ayurveda

Tulasi (Sanskrit:-Surasa) has been used for thousands of years in Ayurveda for its diverse healing properties. It is mentioned in the Charaka Samhita, an ancient Ayurvedic text. *Tulsi* is considered to be an adaptogen, balancing different processes in the body, and helpful for adapting to stress. Marked by its strong aroma and astringent taste, it is regarded in Ayurveda as a kind of "elixir of life" and believed to promote longevity.

Tulasi extracts are used in ayurvedic remedies for a variety of ailments. Traditionally, *tulasi* is taken in many forms: as herbal tea, dried powder, fresh leaf or mixed with *ghee*. Essential oil extracted from Karpoora *tulasi* is mostly used for medicinal purposes and in herbal cosmetics.

Thai Cuisine

The leaves of holy basil, known as *kaphrao* in the Thai language, are commonly used in Thai cuisine. *Kaphrao* should not be confused with *horapha*, which is normally known as Thai basil, or with Thai lemon basil.

Phat kaphrao mu – holy basil with pork

The best-known dish made with this herb is *phat kaphrao* — a stir-fry of Thai holy basil with meats, seafood or, as in *khao phat kraphao*, with rice.

Insect Repellent

For centuries, the dried leaves have been mixed with stored grains to repel insects. In Sri Lanka this plant is used as a mosquito repellent. Sinhala: Maduruthalaa

Chemical Composition

Some of the active chemical constituents of *tulsi* are: oleanolic acid, ursolic acid, rosmarinic acid,eugenol, carvacrol, linalool, β-caryophyllene (about 8%),

Tulsi essential oil has been found to consist mostly of eugenol (~70%) β-elemene (~11.0%), β-caryophyllene (~8%) and germacrene (~2%), with the balance being made up of various trace compounds, (mostly terpenes).

Isolated *O. sanctum* extracts have some antibacterial activity against *E. coli*, *S. aureus* and *P. aeruginosa*.

Genome Sequence

The genome of Tulsi plant has been sequenced and the draft genome has been published independently by research teams from CSIR-Central Institute of Medicinal and Aromatic Plants at Lucknow and National Centre for Biological Sciences at Bengaluru. The genome size was estimated to be 612 mega bases and results from the sequencing project show that certain metabolite-biosynthesis genes such as genes for biosynthesis of Anthocyanin in *Krishna Tulsi* variety, Ursolic acid and Eugenol in *Rama Tulsi* variety were expressed in large quantities. These metabolites were shown to have anti-cancerous properties as well. It was further commented that these metabolites could be utilized as anti-cancerous drugs.

Aloe

Aloe also written *Aloë*, is a genus containing over 500 species of flowering succulent plants. The most widely known species is *Aloe vera*, or "true aloe", so called because it is cultivated as the standard source of so-called "aloe vera" for assorted pharmaceutical purposes. Other species, such as *Aloe ferox*, also are cultivated or harvested from the wild for similar applications.

Succulent plants, such as this aloe, store water in their enlarged fleshy leaves, stems, or roots, as shown in this split aloe leaf. This allows them to survive in arid environments.

Aloe vossii

The APG IV system (2016) places the genus in the family Asphodelaceae, subfamily Asphodeloideae. In the past, it has been assigned to the family Aloaceae (now included in the Asphodelaceae) or to a broadly circumscribed family Liliaceae (the lily family). The plant *Agave americana*, which is sometimes called "American aloe", belongs to the Asparagaceae, a different family.

The genus is native to tropical and southern Africa, Madagascar, Jordan, the Arabian Peninsula, and various islands in the Indian Ocean (Mauritius, Réunion, Comoros, etc.). A few species have also become naturalized in other regions (Mediterranean, India, Australia, North and South America, etc.).

Description

Most *Aloe* species have a rosette of large, thick, fleshy leaves. *Aloe* flowers are tubular, frequently yellow, orange, pink, or red, and are borne, densely clustered and pendant, at the apex of simple or branched, leafless stems. Many species of *Aloe* appear to be stemless, with the rosette growing directly at ground level; other varieties may have a branched or unbranched stem from which the fleshy leaves spring. They vary in color from grey to bright-green and are sometimes striped or mottled. Some aloes native to South Africa are tree-like (arborescent).

Systematics

The APG IV system (2016) places the genus in the family Asphodelaceae, subfamily Asphodeloideae. In the past it has also been assigned to the families Liliaceae and Aloeaceae, as well as the family Asphodelaceae sensu stricto, before this was merged into the Asphodelaceae sensu lato.

The circumscription of the genus has varied widely. Many genera, such as *Lomatophyllum*, have been brought into synonymy. Species at one time placed in *Aloe*, such as *Agave americana*, have been moved to other genera.

Species

Over 500 species are accepted in the genus *Aloe*, plus even more synonyms and unresolved species, subspecies, varieties, and hybrids. Some of the accepted species are:

- *Aloe aculeata* Pole-Evans
- *Aloe africana* Mill.
- *Aloe albida* (Stapf) Reynolds
- *Aloe albiflora* Guillaumin
- *Aloe arborescens* Mill.
- *Aloe arenicola* Reynolds
- *Aloe argenticauda* Merxm. & Giess
- *Aloe aristata* Haw.
- *Aloe bakeri* Scott-Elliot
- *Aloe ballii* Reynolds
- *Aloe ballyi* Reynolds
- *Aloe barberae* Dyer
- *Aloe brevifolia* Mill.
- *Aloe broomii* Schönland
- *Aloe buettneri* A.Berger
- *Aloe camperi* Schweinf.
- *Aloe capitata* Baker
- *Aloe ciliaris* Haw.

- *Aloe commixta* A.Berger
- *Aloe comosa* Marloth & A.Berger
- *Aloe cooperi* Baker
- *Aloe corallina* Verd.
- *Aloe decumbens* (Reynolds) van Jaarsv.
- *Aloe dewinteri* Giess ex Borman & Hardy
- *Aloe dichotoma* Masson
- *Aloe dinteri* A.Berger
- *Aloe eminens* Reynolds & Bally
- *Aloe erinacea* D.S.Hardy
- *Aloe excelsa* A.Berger
- *Aloe ferox* Mill.
- *Aloe forbesii* Balf.f.
- *Aloe gracilis* Haw.
- *Aloe haemanthifolia* Marloth & A.Berger
- *Aloe helenae* Danguy
- *Aloe hereroensis* Engl.
- *Aloe inermis* Forssk.
- *Aloe inyangensis* Christian
- *Aloe jawiyon* S.J.Christie, D.P.Hannon & Oakman ex A.G.Mill.
- *Aloe jucunda* Reynolds
- *Aloe juddii* van Jaarsv.
- *Aloe khamiesensis* Pillans
- *Aloe kilifiensis* Christian
- *Aloe maculata* All.
- *Aloe marlothii* A.Berger
- *Aloe namibensis* Giess
- *Aloe nyeriensis* Christian & I.Verd.
- *Aloe pearsonii* Schönland
- *Aloe peglerae* Schönland

- *Aloe perfoliata* L.
- *Aloe perryi* Baker
- *Aloe petricola* Pole-Evans
- *Aloe pillansii* L.Guthrie
- *Aloe plicatilis* (L.) Mill.
- *Aloe polyphylla* Pillans
- *Aloe ramosissima* Pillans
- *Aloe rauhii* Reynolds
- *Aloe reynoldsii* Letty
- *Aloe scobinifolia* Reynolds & Bally
- *Aloe sinkatana* Reynolds
- *Aloe sladeniana* Pole-Evans
- *Aloe squarrosa* Baker ex Balf.f.
- *Aloe striata* Haw.
- *Aloe striatula* Haw.
- *Aloe succotrina* Lam.
- *Aloe suzannae* Decary
- *Aloe tenuior* Haw.
- *Aloe thraskii* Baker
- *Aloe variegata* L.
- *Aloe vera* (L.) Burm.f.
- *Aloe viridiflora* Reynolds
- *Aloe wildii* (Reynolds) Reynolds

Multiple *Aloe* species with a variety of growth forms.

In addition to the species and hybrids between species within the genus, several hybrids with other genera have been created in cultivation, such as between *Aloe* and *Gasteria*, and between *Aloe* and *Astroloba*.

Uses

Aloe species are frequently cultivated as ornamental plants both in gardens and in pots. Many aloe species are highly decorative and are valued by collectors of succulents. *Aloe vera* is used both internally and externally on humans, and is claimed to have some medicinal effects, which have been supported by scientific and medical research. They can also be made into types of special soaps.

Historical Uses

Depiction of Aloe, labeled in Greek (Aloë) from the Juliana Anicia Codex, a copy, written in Constantinople in 515 AD, of Dioscorides' 1st century AD work.

Historical use of various aloe species is well documented. Documentation of the clinical effectiveness is available, although relatively limited.

Of the 500+ species, only a few were used traditionally as herbal medicines, *Aloe vera* again being the most commonly used species. Also included are *A. perryi* and *A. ferox*. The Ancient Greeks and Romans used *Aloe vera* to treat wounds. In the Middle Ages, the yellowish liquid found inside the leaves was favored as a purgative. Unprocessed aloe that contains aloin is generally used as a laxative, whereas processed juice does not usually contain significant aloin.

Some species, particularly *Aloe vera*, are used in alternative medicine and first aid. Both the translucent inner pulp and the resinous yellow aloin from wounding the aloe plant are used externally for skin discomforts. As an herbal medicine, *Aloe vera* juice is commonly used internally for digestive discomfort.

Relatively few studies about possible benefits of aloe gel taken internally have been conducted. Components of *Aloe* have shown the possibility of inhibiting tumor growth in animal studies, but these effects have not been demonstrated clinically in humans. Some studies in animal models indicate that extracts of *Aloe* have a significant antihyperglycemic effect, and may be useful in treating Type II diabetes, but these studies have not been confirmed in humans.

According to Cancer Research UK, a potentially deadly product called T-UP is made of concentrated aloe, and promoted as a cancer cure. They say "there is currently no evidence that aloe products can help to prevent or treat cancer in humans".

Aloin in OTC Laxative Products

On May 9, 2002, the US Food and Drug Administration issued a final rule banning the use of aloin, the yellow sap of the aloe plant, for use as a laxative ingredient in over-the-counter drug products. Most aloe juices today do not contain significant aloin.

Chemical Properties

According to W. A. Shenstone, two classes of aloins are recognized: (1) nataloins, which yield picric and oxalic acids with nitric acid, and do not give a red coloration with nitric acid; and (2) barbaloins, which yield aloetic acid ($C_7H_2N_3O_5$), chrysammic acid ($C_7H_2N_2O_6$), picric and oxalic acids with nitric acid, being reddened by the acid. This second group may be divided into a-barbaloins, obtained from Barbados *Aloe*, and reddened in the cold, and b-barbaloins, obtained from *Aloe Socotrina* and Zanzibar *Aloe*, reddened by ordinary nitric acid only when warmed or by fuming acid in the cold. Nataloin ($2C_{17}H_{13}O_7 \cdot H_2O$) forms bright-yellow scales, barbaloin ($C_{17}H_{18}O_7$) prismatic crystals. *Aloe* species also contain a trace of volatile oil, to which their odour is due.

Flavoring

Aloe perryi, *A. barbadensis*, *A. ferox*, and hybrids of this species with *A. africana* and *A. spicata* are listed as natural flavoring substances in the US government *Electronic Code of Federal Regulations*. *Aloe Socotrina* is said to be used in yellow Chartreuse.

Heraldic Occurrence

Aloe rubrolutea occurs as a charge in heraldry, for example in the Civic Heraldry of Namibia.

Turmeric

Botanical view of *Curcuma longa*

Turmeric is a rhizomatous herbaceous perennial plant (*Curcuma longa*) of the ginger family, Zingiberaceae. It is native to southern Asia, requiring temperatures between 20 and 30 °C (68 and 86 °F) and a considerable amount of annual rainfall to thrive. Plants are gathered annually for their rhizomes and propagated from some of those rhizomes in the following season.

When not used fresh, the rhizomes are boiled for about 30–45 minutes and then dried in hot ovens, after which they are ground into a deep-orange-yellow powder commonly used as a coloring and flavoring agent in the cuisines of Bangladesh, India, Indonesia, Iran and Pakistani cuisine, especially for curries, as well as for dyeing.

Although long-used in Ayurvedic medicine to treat various diseases, there is little high-quality clinical evidence for use of turmeric or its main constituent, curcumin, as a therapy.

History and Etymology

Turmeric has been used in Asia for thousands of years and is a major part of Ayurveda, Siddha medicine, Unani, and traditional Chinese medicine. It was first used as a dye, and then later for its medicinal properties.

The origin of the name is uncertain, possibly deriving from Middle English or Early Modern English as *turmeryte* or *tarmaret*. Speculation exists that it may be of Latin origin, *terra merita* ("meritorious earth"). The name of the genus, *Curcuma*, is from an Arabic name of both saffron and turmeric. In Sanskrit, turmeric is called Haridraa. It is paired with vermillion in auspicious events and religious ceremonies. In Kannada, turmeric is called Arishina.

Botanical Description

Appearance

Turmeric is a perennial herbaceous plant that reaches up to 1 m (3 ft 3 in) tall. Highly branched, yellow to orange, cylindrical, aromatic rhizomes are found. The leaves are alternate and arranged in two rows. They are divided into leaf sheath, petiole, and leaf blade. From the leaf sheaths, a false stem is formed. The petiole is 50 to 115 cm (20 to 45 in) long. The simple leaf blades are usually 76 to 115 cm (30 to 45 in) long and rarely up to 230 cm (91 in). They have a width of 38 to 45 cm (15 to 18 in) and are oblong to elliptic, narrowing at the tip.

Inflorescence, Flower, and Fruit

Turmeric flower

In China, the flowering time is usually in August. Terminally on the false stem is a 12 to 20 cm (4.7 to 7.9 in) long inflorescence stem containing many flowers. The bracts are light green and ovate to oblong with a blunt upper end with a length of 3 to 5 cm (1.2 to 2.0 in).

Wild turmeric, Australia

At the top of the inflorescence, stem bracts are present on which no flowers occur; these are white to green and sometimes, tinged reddish-purple, and the upper ends are tapered.

The hermaphrodite flowers are zygomorphic and threefold. The three 0.8 to 1.2 cm (0.3 to 0.5 in) long sepals are fused, white, have fluffy hairs and the three calyx teeth are unequal. The three bright-yellow petals are fused into a corolla tube up to 3 cm (1.2 in) long. The three corolla lobes have a length of 1.0 to 1.5 cm (0.39 to 0.59 in) and are triangular with soft-spiny upper ends. While the average corolla lobe is larger than the two lateral, only the median stamen of the inner circle is fertile. The dust bag is spurred at its base. All other stamens are converted to staminodes. The outer staminodes are shorter than the labellum. The labellum is yellowish, with a yellow ribbon in its center and it is obovate, with a length from 1.2 to 2.0 cm (0.47 to 0.79 in). Three carpels are under a constant, trilobed ovary adherent, which is sparsely hairy. The fruit capsule opens with three compartments.

Biochemical Composition

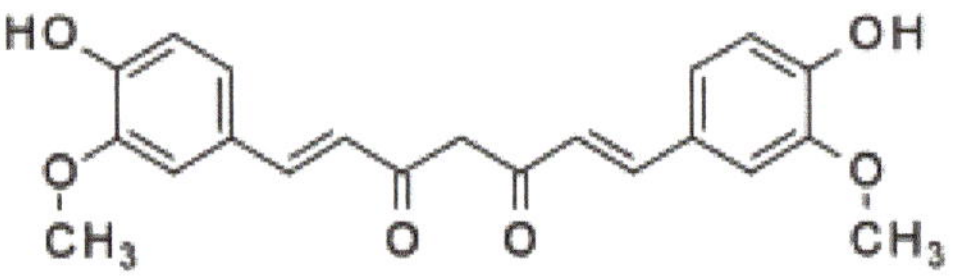

Curcumin keto form

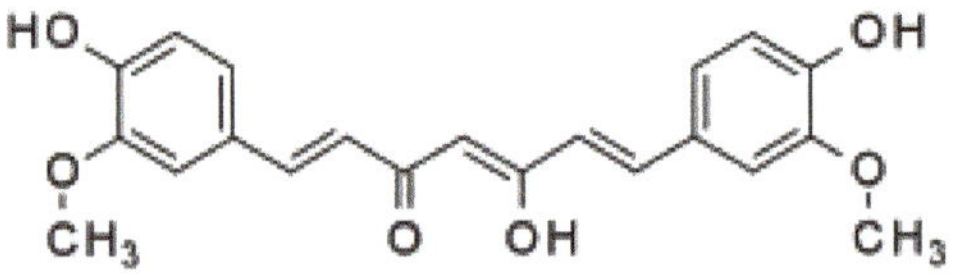

Curcumin enol form

The most important chemical components of turmeric are a group of compounds called curcuminoids, which include curcumin (diferuloylmethane), demethoxycurcumin, and bisdemethoxycurcumin. The best-studied compound is curcumin, which constitutes 3.14% (on average) of powdered turmeric, however, there are big variations in curcumin content in the different lines of the species Curcuma longa (1–3189 mg/100 g). In addition, other important volatile oils include turmerone, atlantone, and zingiberene. Some general constituents are sugars, proteins, and resins.

Uses

Traditional Medicine

Turmeric grows wild in the forests of South and Southeast Asia where it is collected for use in Indian traditional medicine (also called Siddha or Ayurveda). From clinical research, there is no high-quality evidence that turmeric has medicinal properties.

Culinary

Turmeric powder

Turmeric rhizome and powder

Curry using turmeric, referred to as *haldi ki Sabji*, a dish from India

Turmeric is one of the key ingredients in many Asian dishes. Its use as a coloring agent is not of primary value in South Asian cuisine.

Turmeric is used mostly in savory dishes, but also is used in some sweet dishes, such as the cake *sfouf*. In India, turmeric plant leaf is used to prepare special sweet dishes, *patoleo*, by layering rice flour and coconut-jaggery mixture on the leaf, then closing and steaming it in a special copper steamer (*goa*).

In recipes outside South Asia, turmeric sometimes is used as an agent to impart a golden yellow color. It is used in canned beverages, baked products, dairy products, ice cream, yogurt, yellow cakes, orange juice, biscuits, popcorn color, cereals, sauces, gelatins, etc. It is a significant ingredient in most commercial curry powders.

Most turmeric is used in the form of rhizome powder. In some regions (especially in Maharashtra, Goa, Konkan, and Kanara), turmeric leaves are used to wrap and cook food. Turmeric leaves are mainly used in this way in areas where turmeric is grown locally, since the leaves used are freshly picked. Turmeric leaves impart a distinctive flavor.

Although typically used in its dried, powdered form, turmeric also is used fresh, like ginger. It has numerous uses in East Asian recipes, such as pickle that contains large chunks of soft turmeric, made from fresh turmeric.

Turmeric is used widely as a spice in South Asian and Middle Eastern cooking. Many Persian dishes use turmeric as a starter ingredient. Various Iranian *khoresh* dishes are started using onions caramelized in oil and turmeric, followed by other ingredients. The Moroccan spice mix ras el hanout typically includes turmeric.

In India and Nepal, turmeric is widely grown and extensively used in many vegetable and meat dishes for its color. It also is used in Nepal for its supposed value in traditional medicine.

In South Africa, turmeric is used to give boiled white rice a golden colour.

In Vietnamese cuisine, turmeric powder is used to color and enhance the flavors of certain dishes, such as *bánh xèo, bánh khọt*, and *mi quang*. The powder is used in many other Vietnamese stir-fried and soup dishes.

The staple Cambodian curry paste *kroeung*, used in many dishes including *Amok*, typically contains fresh turmeric.

In Indonesia, turmeric leaves are used for Minang or Padang curry base of Sumatra, such as *rendang, sate padang*, and many other varieties.

In Thailand, fresh turmeric rhizomes are used widely in many dishes, in particular in the southern Thai cuisine, such as the yellow curry and turmeric soup.

In medieval Europe, turmeric became known as Indian saffron because it was used widely as an alternative to the far more expensive saffron spice.

Dye

Turmeric makes a poor fabric dye, as it is not very light fast, but is commonly used in Indian and Bangladeshi clothing, such as saris and Buddhist monks's robes. Turmeric (coded as E100, when used as a food additive), is used to protect food products from sunlight. The oleoresin is used for oil-containing products. A curcumin and polysorbate solution or curcumin powder dissolved in alcohol is used for water-containing products. Over-coloring, such as in pickles, relishes, and mustard, is sometimes used to compensate for fading.

In combination with annatto (E160b), turmeric has been used to color cheeses, yogurt, dry mixes, salad dressings, winter butter, and margarine. Turmeric also is used to give a yellow color to some prepared mustards, canned chicken broths, and other foods (often as a much cheaper replacement for saffron).

Indicator

Turmeric paper, also called curcuma paper or in German literature, *Curcumapapier*, is paper steeped in a tincture of turmeric and allowed to dry. It is used in chemical analysis as an indicator for acidity and alkalinity. The paper is yellow in acidic and neutral solutions and turns brown to reddish-brown in alkaline solutions, with transition between pH of 7.4 and 9.2.

For pH detection, turmeric paper has been replaced in common use by litmus paper. Turmeric may be used as a substitute for phenolphthalein, as its color change pH range is similar.

Traditional Uses

Curcuma domestica Valeton, a drawing by A. Bernecker around 1860

In Ayurvedic practices, turmeric has been used as an attempted treatment for a variety of internal disorders, such as indigestion, throat infections, common colds, or liver ailments, as well as topically, to cleanse wounds or treat skin sores.

Turmeric is considered auspicious and holy in India and has been used in various Hindu ceremonies for millennia. It remains popular in India for wedding and religious ceremonies.

The plant is used in Poosai (Tamil) to represent a form of the Tamil Goddess Kottravai.

In Eastern India, the plant is used as one of the nine components of navapatrika along with young plantain or banana plant, taro leaves, barley (*jayanti*), wood apple (*bilva*), pomegranate (*darimba*), *asoka*, *manaka* or *manakochu*, and rice paddy. The Navapatrika worship is an important part of Durga festival rituals.

It is used in *poosai* to make a form of Ganesha. Yaanaimugathaan, the remover of obstacles, is invoked at the beginning of almost any ceremony. A form of Yaanaimugathaan for this purpose is made by mixing turmeric with water and forming it into a cone-like shape.

Haldi ceremony (called *Gaye holud* in Bengal) (literally "yellow on the body") is a ceremony observed during Hindu wedding celebrations in many parts of India including Bengal, Punjab, Maharashtra, and Gujarat. The 'ceremony takes place one or two days before the religious and legal Bengali wedding ceremonies. The turmeric paste is applied by friends to the bodies of the couple. This is said to soften the skin, but also colors them with the distinctive yellow hue that gives its name to this ceremony. It may be a joint or separate event for the families of the bride and groom. During the Tamil festival *Pongal*, a whole turmeric plant with fresh rhizomes is offered as a thanksgiving offering to Suryan, the sun deity. Also, the fresh plant sometimes is tied around the Pongal pot in which an offering of *pongal* is prepared.

In Tamil Nadu and Andhra Pradesh, as a part of the Tamil/Telugu marriage ritual, dried turmeric tuber tied with string is used to create a Thali necklace, the equivalent of marriage rings in western cultures. In western and coastal India, during weddings of the Marathi and Konkani people, Kannada Brahmins turmeric tubers are tied with strings by the couple to their wrists during a ceremony, *Kankanabandhana*.

Friedrich Ratzel reported in *The History of Mankind* during 1896, that in Micronesia, turmeric powder was applied for embellishment of body, clothing, utensils, and ceremonial uses.

Adulteration

As turmeric and other spices are commonly sold by weight, the potential exists for powders of toxic, cheaper agents with a similar color to be added, such as lead(II,IV) oxide, giving turmeric an orange-red color instead of its native gold-yellow. Another common adulterant in turmeric, metanil yellow (also known as acid yellow 36), is considered an illegal dye for use in foods by the British Food Standards Agency.

Research

Claims that curcumin in turmeric may help to reduce inflammation have not been supported by strong studies.

Turmeric or its principal constituent, curcumin, has been studied in numerous clinical trials for various human diseases and conditions, but the conclusions have either been uncertain or negative.

Phyllanthus Emblica

Phyllanthus emblica, also known as emblic, emblic myrobalan, myrobalan, Indian gooseberry, Malacca tree, or amla from Sanskrit amalaki is a deciduous tree of the family Phyllanthaceae. It is known for its edible fruit of the same name.

Plant Morphology and Harvesting

The tree is small to medium in size, reaching 1–8 m (3 ft 3 in–26 ft 3 in) in height. The branchlets are not glabrous or finely pubescent, 10–20 cm (3.9–7.9 in) long, usually deciduous; the leaves are simple, subsessile and closely set along branchlets, light green, resembling pinnate leaves. The flowers are greenish-yellow. The fruit is nearly spherical, light greenish yellow, quite smooth and hard on appearance, with six vertical stripes or furrows.

Ripening in autumn, the berries are harvested by hand after climbing to upper branches bearing the fruits. The taste of Indian emblic is sour, bitter and astringent, and it is quite fibrous. In India, it is common to eat emblic steeped in salt water and red chilli powder to make the sour fruits palatable.

Cultural and Religious Significance

Fruit with young leaves and flower buds.

The tree is considered sacred by Hindus as God Vishnu is believed to dwell in it. The tree is worshiped on Amalaka Ekadashi.

In other Hindu beliefs, amla is said to have originated from the drops of Amrit which spilled on earth accidentally, because of the fight of gods and demons after ksheera sagar manthan. This religious belief makes claims that it almost cures every disease and is also good in extending the longevity of life.

In the Sanskrit Buddhist tradition, half an amalaka fruit was the final gift to the Buddhist sangha by the great Indian emperor Ashoka. This is illustrated in the Ashokavadana in the following verses:

"A great donor, the lord of men, the eminent Maurya Ashoka, has gone from being lord of Jambudvipa [the continent] to being lord of half a myrobalan." (Strong, 1983, p. 99) This deed became so famous that a stupa was created to mark the place of the event in modern-day Patna and was known as the Amalaka stupa.

According to Hindu tradition, Adi Shankara of Kerala composed and recited the Kanakadhara stotram in praise of Mahalakshmi to make a poor Brahmin lady get wealth, in return for a single amla presented to him as bhiksha on an auspicious dwadashi day.

According to a Tamil legend, Avvaiyar, a female poet, ethicist and political activist of the Sangam period was gifted with one amla by King Athiyaman to give her long life.

Amalaka at the top of the Lingaraj temple in Bhubaneswar

In Theravada Buddhism, this plant is said to have been used as the tree for achieving enlightenment, or Bodhi by twenty first Buddha named Phussa Buddha.

In Indian temple architecture, an Amalaka, is a stone disk, usually with ridges on the rim, that sits atop a temple's main tower (Shikhara). The shape of the amalaka is thought to have been inspired by the fruit of the amla tree.

Traditional Uses

Emblic seller in Pollachi, India

Traditional Medicine

In traditional Indian medicine, dried and fresh fruits of the plant are used. All parts of the plant are used in various Ayurvedic/Unani medicine (*Jawarish amla*) herbal preparations, including the fruit, seed, leaves, root, bark and flowers. According to Ayurveda, amla fruit is sour (*amla*) and astringent (*kashaya*) in taste (*rasa*), with sweet (*madhura*), bitter (*tikta*) and pungent (*katu*) secondary tastes (*anurasas*). Its qualities (*gunas*) are light (*laghu*) and dry (*ruksha*), the postdigestive effect (*vipaka*) is sweet (*madhura*) and its energy (*virya*) is cooling (*shita*).

According to Ayurveda, amla balances all three doshas. While amla is unusual in that it contains five out of the six tastes recognized by Ayurveda, it is most important to recognize the effects of the "virya", or potency, and "vipaka", or post-digestive effect. Considered in this light, amla is particularly helpful in reducing *pitta* because of its cooling energy. It also balances both Pitta and *vata* by virtue of its sweet taste. The *kapha* is balanced primarily due to its drying action. It may be used as a *rasayana* (rejuvenative) to promote longevity, and traditionally to enhance digestion (*dipanapachana*), treat constipation (*anuloma*), reduce fever (*jvaraghna*), purify the blood (*raktaprasadana*), reduce cough (*kasahara*), alleviate asthma (*svasahara*), strengthen the heart (*hrdaya*), benefit the eyes (*chakshushya*), stimulate hair growth (*romasanjana*), enliven the body (*jivaniya*), and enhance intellect (*medhya*).

In Ayurvedic polyherbal formulations, Indian gooseberry is a common constituent, and most notably is the primary ingredient in an ancient herbal *rasayana* called *Chyawanprash*. This formula, which contains 43 herbal ingredients as well as clarified butter, sesame oil, sugar cane juice, and honey, was first mentioned in the Charaka Samhita as a premier rejuvenative compound.

Culinary use

Maharashtra is one of the largest producers and suppliers of Indian gooseberries. In this region the fruit is commonly pickled with salt, oil, and spices. The amla fruit is eaten raw or cooked into various dishes. In Andhra Pradesh, tender varieties are used to prepare *dal* (a lentil preparation), and *amle ka murabbah*, a sweet dish made by soaking the berries in sugar syrup until they are candied. It is traditionally consumed after meals.

In the Batak area of Sumatra, Indonesia, the inner bark is used to impart an astringent, bitter taste to the broth of a traditional fish soup known as *holat*.

Other uses

Popularly used in inks, shampoos and hair oils, the high tannin content of Indian gooseberry fruit serves as a mordant for fixing dyes in fabrics. Amla shampoos and hair oil are traditionally believed to nourish the hair and scalp and prevent premature grey hair.

Research

In vitro studies of *P. emblica* extracts on osteoclasts have been done to establish their potential biological effects. Experimental preparations of leaves, bark or fruit are under research on mechanisms of inflammation and alanine transaminase activity.

Chemical Constituents

Although these fruits are reputed to contain high amounts of ascorbic acid (vitamin C), up to 445 mg per 100 g, the specific contents are disputed, and the overall antioxidant strength of amla may derive instead from its high density of ellagitannins such as emblicanin A (37%), emblicanin B (33%), punigluconin (12%) and pedunculagin (14%). It also contains punicafolin and phyllanemblinin A, phyllanemblin other polyphenols: flavonoids, kaempferol, ellagic acid and gallic acid.

Datura Stramonium

Datura stramonium, known by the common names jimsonweed or Devil's snare, is a plant in the nightshade family. It is believed to have originated in Mexico, but has now become naturalized in many other regions. Other common names for D. stramonium include thornapple and moon flower, and it has the Spanish name toloache. Other names for the plant include hell's bells, devil's trumpet, devil's weed, tolguacha, Jamestown weed, stinkweed, locoweed, pricklyburr, and devil's cucumber.

Datura has been used in traditional medicine to relieve asthma symptoms and as an analgesic during surgery or bonesetting. It is also a powerful hallucinogen and deliriant, which is used entheogenically for the intense visions it produces. However, the tropane alkaloids responsible for both the medicinal and hallucinogenic properties are fatally toxic in only slightly higher amounts than the medicinal dosage, and careless use often results in hospitalizations and deaths.

Description

Datura stramonium is a foul-smelling, erect, annual, freely branching herb that forms a bush up to 60 to 150 cm (2 to 5 ft) tall.

The root is long, thick, fibrous, and white. The stem is stout, erect, leafy, smooth, and pale yellow-green. The stem forks off repeatedly into branches, and each fork forms a leaf and a single, erect flower.

The leaves are about 8 to 20 cm (3–8 in) long, smooth, toothed, soft, and irregularly undulated. The upper surface of the leaves is a darker green, and the bottom is a light green. The leaves have a bitter and nauseating taste, which is imparted to extracts of the herb, and remains even after the leaves have been dried.

Datura stramonium generally flowers throughout the summer. The fragrant flowers are trumpet-shaped, white to creamy or violet, and 6 to 9 cm ($2\frac{1}{2}$–$3\frac{1}{2}$ in) long, and grow on short stems from either the axils of the leaves or the places where the branches fork. The calyx is long and tubular, swollen at the bottom, and sharply angled, surmounted by five sharp teeth. The corolla, which is folded and only partially open, is white, funnel-shaped, and has prominent ribs. The flowers open at night, emitting a pleasant fragrance, and are fed upon by nocturnal moths.

The egg-shaped seed capsule is 3 to 8 cm (1–3 in) in diameter and either covered with spines or bald. At maturity, it splits into four chambers, each with dozens of small, black seeds.

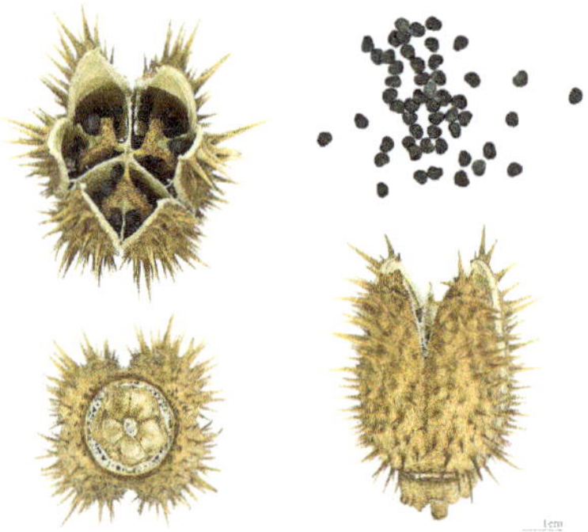

Fruits and seeds – MHNT

Range and Habitat

Datura stramonium is native to North America, but was spread to the Old World early. It was scientifically described and named by Swedish botanist Carl Linnaeus in 1753, although it had been described a century earlier by botanists such as Nicholas Culpeper. Today, it grows wild in all the world's warm and moderate regions, where it is found along roadsides and at dung-rich livestock enclosures. In Europe, it is found as a weed on wastelands and in garbage dumps.

The seed is thought to be carried by birds and spread in their droppings. Its seeds can lie dormant underground for years and germinate when the soil is disturbed. People who discover it growing in their gardens, and are worried about its toxicity, have been advised to dig it up or have it otherwise removed.

Toxicity

All parts of *Datura* plants contain dangerous levels of the tropane alkaloids atropine, hyoscyamine, and scopolamine, which are classified as deliriants, or anticholinergics. The risk of fatal overdose is high among uninformed users, and many hospitalizations occur amongst recreational users who ingest the plant for its psychoactive effects.

The amount of toxins varies widely from plant to plant. As much as a 5:1 variation can be found between plants, and a given plant's toxicity depends on its age, where it is growing, and the local weather conditions. Additionally, within a given datura plant, toxin concentration varies by part and even from leaf to leaf. When the plant is younger, the ratio of scopolamine to atropine is about 3:1; after flowering, this ratio is reversed, with the amount of scopolamine continuing to decrease as the plant gets older. In traditional cultures, a great deal of experience with and detailed knowledge of *Datura* was critical to minimize harm. An individual datura seed contains about 0.1 mg of atropine, and the approximate fatal dose for adult humans is >10 mg atropine or >2–4 mg scopolamine.

Datura intoxication typically produces delirium (as contrasted to hallucination), hyperthermia, tachycardia, bizarre behavior, and severe mydriasis with resultant painful photophobia that can last several days. Pronounced amnesia is another commonly reported effect. The onset of symptoms generally occurs around 30 to 60 minutes after ingesting the herb. These symptoms generally last from 24 to 48 hours, but have been reported in some cases to last as long as two weeks.

As with other cases of anticholinergic poisoning, intravenous physostigmine can be administered in severe cases as an antidote.

Use in Traditional Medicine

D. stramonium var. *tatula*, flower (front)

In Ayurveda, datura has long been used for asthma symptoms. The active agent is atropine. The leaves are generally smoked either in a cigarette or a pipe. During the late 18th century, James Anderson, the English Physician General of the East India Company, learned of the practice and popularized it in Europe.

John Gerard's *Herball* (1597) states,

The juice of Thornapple, boiled with hog's grease, cureth all inflammations whatsoever, all manner of burnings and scaldings, as well of fire, water, boiling lead, gunpowder, as that which comes by lightning and that in very short time, as myself have found in daily practice, to my great credit and profit.

The Zuni people once used datura as an analgesic to render patients unconscious while broken bones were set. The Chinese also used it as a form of anesthesia during surgery.

Spiritual uses

Datura seedpod, opening up to release seeds inside

The ancient inhabitants of what is today central and southern California used to ingest the small black seeds of datura to "commune with deities through visions". Across the Americas, other indigenous peoples such as the Algonquin, Navajo, Cherokee, Luiseño and the indigenous peoples of Marie-Galante also used this plant in sacred ceremonies for its hallucinogenic properties. In Ethiopia, some students and *debtrawoch* (lay priests), use *D. stramonium* to "open the mind" to be more receptive to learning, and creative and imaginative thinking.

In his book, *The Serpent and the Rainbow*, Wade Davis identified *D. stramonium*, called "zombi cucumber" in Haiti, as a central ingredient of the concoction vodou priests use to create zombies.

The common name "datura" has its roots in India, where the plant is considered particularly sacred—believed to be a favorite of Nataraja in Shaivism.

Cultivation

Datura prefers rich, calcareous soil. Adding nitrogen fertilizer to the soil will increase the concentration of alkaloids present in the plant. Datura can be grown from seed, which is sown with several feet between plants. Datura is sensitive to frost, so should be sheltered during cold weather. The plant is harvested when the fruits are ripe, but still green. To harvest, the entire plant is cut down, the leaves are stripped from the plant, and everything is left to dry. When the fruits begin to burst open, the seeds are harvested. For intensive plantations, leaf yields of 1,100 to 1,700 kilograms per hectare (1,000 to 1,500 lb/acre) and seed yields of 780 kg/ha (700 lb/acre) are possible.

Etymology

The genus name is derived from the plant's Hindi name *dhatūra*, ultimately from Sanskrit *dhattūra* 'white thorn-apple'. *Stramonium* is originally from Greek "nightshade" and "mad".

In the United States, the plant is called "jimsonweed", or more rarely "Jamestown weed"; it got this name from the town of Jamestown, Virginia, where British soldiers consumed it while attempting to suppress Bacon's Rebellion. They spent 11 days in altered mental states:

The James-Town Weed (which resembles the Thorny Apple of Peru, and I take to be the plant so call'd) is supposed to be one of the greatest coolers in the world. This being an early plant, was gather'd very young for a boil'd salad, by some of the soldiers sent thither to quell the rebellion of Bacon (1676); and some of them ate plentifully of it, the effect of which was a very pleasant comedy, for they turned natural fools upon it for several days: one would blow up a feather in the air; another would dart straws at it with much fury; and another, stark naked, was sitting up in a corner like a monkey, grinning and making mows [grimaces] at them; a fourth would fondly kiss and paw his companions, and sneer in their faces with a countenance more antic than any in a Dutch droll.

In this frantic condition they were confined, lest they should, in their folly, destroy themselves—though it was observed that all their actions were full of innocence and good nature. Indeed, they were not very cleanly; for they would have wallowed in their own excrements, if they had not been prevented. A thousand such simple tricks they played, and after eleven days returned themselves again, not remembering anything that had passed.

—*Robert Beverley, Jr., The History and Present State of Virginia, Book II: Of the Natural Product and Conveniencies in Its Unimprov'd State, Before the English Went Thither, 1705*

Traditional Chinese Medicine

Traditional Chinese medicine is a style of traditional medicine informed by modern medicine but built on a foundation of more than 2,500 years of Chinese medical practice that includes various forms of herbal medicine, acupuncture, massage (tui na), exercise (qigong), and dietary therapy. It is primarily used as a complementary alternative medicine approach. TCM is widely used in China and is becoming increasingly available in Europe and North America.

One of the basic tenets of TCM "holds that the body's vital energy (*chi* or *qi*) circulates through channels, called *meridians,* that have branches connected to bodily organs and functions." Concepts of the body and of disease used in TCM reflect its ancient origins and its emphasis on dynamic processes over material structure, similar to European humoral theory. Scientific investigation has not found histological or physiological evidence for traditional Chinese concepts such as *qi*, meridians, and acupuncture points. The TCM theory and practice are not based upon scientific knowledge, and there is disagreement between TCM practitioners on what diagnosis and treatments should be used for any given patient. The effectiveness of Chinese herbal medicine remains poorly researched and documented. There are concerns over a number of potentially toxic plants, animal parts, and mineral Chinese medicinals. There are also concerns over illegal trade and transport of endangered species including rhinoceroses and tigers, and the welfare of specially farmed animals including bears. A review of cost-effectiveness research for TCM found that studies had low levels of evidence, but so far have not shown benefit outcomes. Pharmaceutical research has explored the potential for creating new drugs from traditional remedies, with few successful results. A *Nature* editorial described TCM as "fraught with pseudoscience", and said that the most obvious reason why it hasn't delivered many cures is that the majority of its treatments have no logical mechanism of action. Proponents propose that research has so far missed key features of the art of TCM, such as unknown interactions between various ingredients and complex interactive biological systems.

The doctrines of Chinese medicine are rooted in books such as the *Yellow Emperor's Inner Canon* and the *Treatise on Cold Damage*, as well as in cosmological notions such as yin-yang and the five phases. Starting in the 1950s, these precepts were standardized in the People's Republic of China, including attempts to integrate them with modern notions of anatomy and pathology. In the 1950s, the Chinese government promoted a systematized form of TCM.

TCM's view of the body places little emphasis on anatomical structures, but is mainly concerned with the identification of functional entities (which regulate digestion, breathing, aging etc.). While health is perceived as the harmonious interaction of these entities and the outside world, disease is interpreted as a disharmony in interaction. TCM diagnosis aims to trace symptoms to patterns of an underlying disharmony, by measuring the pulse, inspecting the tongue, skin, and eyes, and looking at the eating and sleeping habits of the person as well as many other things.

History

Traces of therapeutic activities in China date from the Shang dynasty (14th–11th centuries BC). Though the Shang did not have a concept of "medicine" as distinct from other fields, their oracular inscriptions on bones and tortoise shells refer to illnesses that affected the Shang royal family: eye disorders, toothaches, bloated abdomen, etc., which Shang elites usually attributed to curses sent by their ancestors. There is no evidence that the Shang nobility used herbal remedies. According to a 2006 overview, the "Documentation of Chinese materia medica (CMM) dates back to around 1,100 BC when only dozens of drugs were first described. By the end of the 16th century, the number of drugs documented had reached close to 1,900. And by the end of the last century, published records of CMM had reached 12,800 drugs."

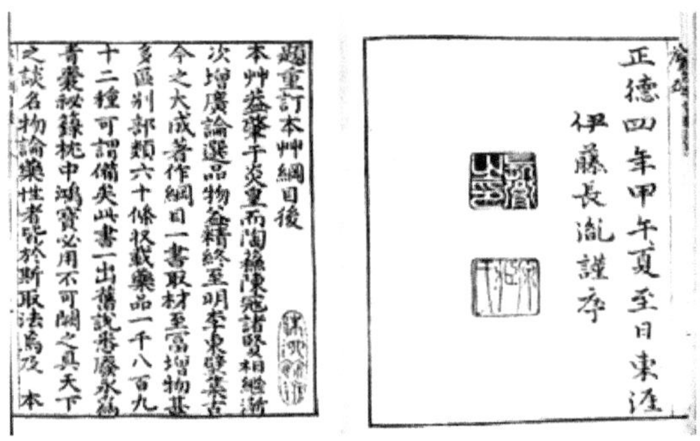

The Compendium of Materia Medica is a pharmaceutical text written by Li Shizhen (1518–1593 AD) during the Ming Dynasty of China. This edition was published in 1593.

Stone and bone needles found in ancient tombs led Joseph Needham to speculate that acupuncture might have been carried out in the Shang dynasty. This being said, most historians now make a distinction between medical lancing (or bloodletting) and acupuncture in the narrower sense of using metal needles to treat illnesses by stimulating specific points along circulation channels ("meridians") in accordance with theories related to the circulation of Qi. The earliest evidence for acupuncture in this sense dates to the second or first century BC.

The *Yellow Emperor's Inner Canon*, the oldest received work of Chinese medical theory, was compiled around the first century BC on the basis of shorter texts from different medical lineages. Written in the form of dialogues between the legendary Yellow Emperor and his ministers, it offers explanations on the relation between humans, their environment, and the cosmos, on the contents of the body, on human vitality and pathology, on the symptoms of illness, and on how to make diagnostic and therapeutic decisions in light of all these factors. Unlike earlier texts like *Recipes for Fifty-Two*

Ailments, which was excavated in the 1970s from a tomb that had been sealed in 168 BC, the *Inner Canon* rejected the influence of spirits and the use of magic. It was also one of the first books in which the cosmological doctrines of Yinyang and the Five Phases were brought to a mature synthesis.

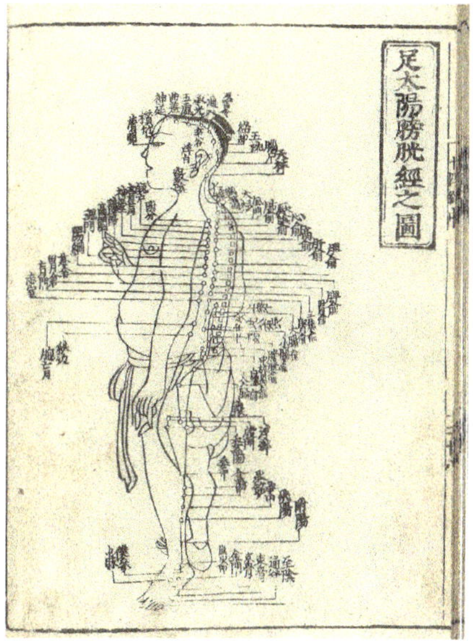

Acupuncture chart from Hua Shou (fl. 1340s, Yuan Dynasty). This image from *Shi si jing fa hui (Expression of the Fourteen Meridians)*. (Tokyo: Suharaya Heisuke kanko, Kyoho gan 1716).

The *Treatise on Cold Damage Disorders and Miscellaneous Illnesses* was collated by Zhang Zhongjing sometime between 196 and 220 CE; at the end of the Han dynasty. Focusing on drug prescriptions rather than acupuncture, it was the first medical work to combine Yinyang and the Five Phases with drug therapy. This formulary was also the earliest public Chinese medical text to group symptoms into clinically useful "patterns" (*zheng*) that could serve as targets for therapy. Having gone through numerous changes over time, the formulary now circulates as two distinct books: the *Treatise on Cold Damage Disorders* and the *Essential Prescriptions of the Golden Casket*, which were edited separately in the eleventh century, under the Song dynasty.

In the centuries that followed the completion of the *Yellow Emperor's Inner Canon*, several shorter books tried to summarize or systematize its contents. The *Canon of Problems* (probably second century CE) tried to reconcile divergent doctrines from the *Inner Canon* and developed a complete medical system centered on needling therapy. The *AB Canon of Acupuncture and Moxibustion* (*Zhenjiu jiayi jing*, compiled by Huangfu Mi sometime between 256 and 282 CE) assembled a consistent body of doctrines concerning acupuncture; whereas the *Canon of the Pulse* (*Maijing*; ca. 280) presented itself as a "comprehensive handbook of diagnostics and therapy."

In 1950, Chairman Mao Zedong made a speech in support of traditional Chinese medicine (TCM) which was influenced by political necessity. Zedong believed he and the Chinese Communist Party should promote TCM but he did not personally believe in TCM and he did not use it. In 1952, the president of the Chinese Medical Association said that, "This One Medicine, will possess a basis in modern natural sciences, will have absorbed the ancient and the new, the Chinese and the foreign, all medical achievements—and will be China's New Medicine!"

Historical Physicians

These include Zhang Zhongjing, Hua Tuo, Sun Simiao, Tao Hongjing, Zhang Jiegu, and Li Shizhen.

Philosophical Background

Traditional Chinese medicine (TCM) is a broad range of medicine practices sharing common concepts which have been developed in China and are based on a tradition of more than 2,000 years, including various forms of herbal medicine, acupuncture, massage (Tui na), exercise (qigong), and dietary therapy. It is primarily used as a complementary alternative medicine approach. TCM is widely used in China and it is also used in the West. Its philosophy is based on Yinyangism (i.e., the combination of Five Phases theory with Yin-yang theory), which was later absorbed by Daoism.

Yin and yang symbol for balance. In Traditional Chinese Medicine, good health is believed to be achieved by a balance between yin and yang.

Yin and Yang

Yin and yang are ancient Chinese concepts which can be traced back to the Shang dynasty (1600–1100 BC). They represent two abstract and complementary aspects that every phenomenon in the universe can be divided into. Primordial analogies for these aspects are the sun-facing (yang) and the shady (yin) side of a hill. Two other commonly used representational allegories of yin and yang are water and fire. In the ying-yang theory, detailed attributions are made regarding the yin or yang character of things:

Phenomenon	**Yin**	**Yang**
Celestial bodies	moon	sun
Gender	female	male
Location	inside	outside
Temperature	cold	hot
Direction	downward	upward
Degree of humidity	damp/moist	dry

The concept of yin and yang is also applicable to the human body; for example, the upper part of the body and the back are assigned to yang, while the lower part of the body are believed to have the yin character. Yin and yang characterization also extends to the various body functions, and – more importantly – to disease symptoms (e.g., cold and heat sensations are assumed to be yin and yang symptoms, respectively). Thus, yin and yang of the body are seen as phenomena whose lack (or over-abundance) comes with characteristic symptom combinations:

- Yin vacuity (also termed "vacuity-heat"): heat sensations, possible night sweats, insomnia, dry pharynx, dry mouth, dark urine, a red tongue with scant fur, and a "fine" and rapid pulse.
- Yang vacuity ("vacuity-cold"): aversion to cold, cold limbs, bright white complexion, long voidings of clear urine, diarrhea, pale and enlarged tongue, and a slightly weak, slow and fine pulse.

TCM also identifies drugs believed to treat these specific symptom combinations, i.e., to reinforce yin and yang.

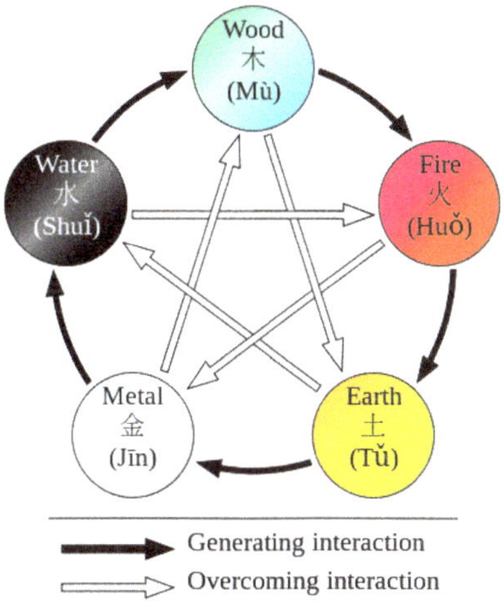

Interactions of Wu Xing

Five Phases Theory

Five Phases (pinyin: *wǔ xíng*), sometimes also translated as the "Five Elements" theory, presumes that all phenomena of the universe and nature can be broken down into five elemental qualities – represented by wood (pinyin: *mù*), fire (pinyin: *huǒ*), earth (pinyin: *tǔ*), metal (pinyin: *jīn*), and water (pinyin: *shuǐ*). In this way, lines of correspondence can be drawn:

Phenomenon	Wood	Fire	Earth	Metal	Water
Direction	East	South	Centre	West	North
Colour	green/blue	red	yellow	white	black
Climate	wind	heat	damp	dryness	cold
Taste	sour	bitter	sweet	acrid	salty
Zang Organ	Liver	Heart	Spleen	Lung	Kidney
Fu Organ	Gallbladder	Small intestine	Stomach	Large intestine	Bladder
Sense organ	eye	tongue	mouth	nose	ears
Facial part	above bridge of nose	between eyes, lower part	bridge of nose	between eyes, middle part	cheeks (below cheekbone)
Eye part	iris	inner/outer corner of the eye	upper and lower lid	sclera	pupil

Strict rules are identified to apply to the relationships between the Five Phases in terms of sequence, of acting on each other, of counteraction, etc. All these aspects of Five Phases theory constitute the basis of the zàng-fŭ concept, and thus have great influence regarding the TCM model of the body. Five Phase theory is also applied in diagnosis and therapy.

Correspondences between the body and the universe have historically not only been seen in terms of the Five Elements, but also of the "Great Numbers" (pinyin: *dà shū*) For example, the number of acu-points has at times been seen to be 365, corresponding with the number of days in a year; and the number of main meridians–12–has been seen as corresponding with the number of rivers flowing through the ancient Chinese empire.

Model of the Body

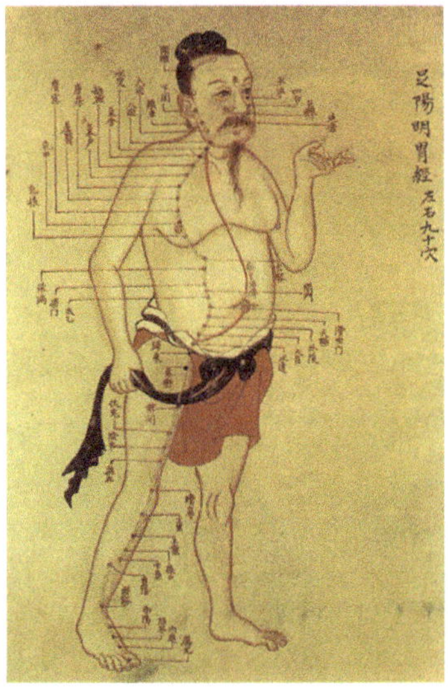

Old Chinese medical chart on acupuncture meridians

TCM "holds that the body's vital energy (*chi* or *qi*) circulates through channels, called *meridians*, that have branches connected to bodily organs and functions." Its view of the human body is only marginally concerned with anatomical structures, but focuses primarily on the body's *functions* (such as digestion, breathing, temperature maintenance, etc.):

"The tendency of Chinese thought is to seek out dynamic functional activity rather than to look for the fixed somatic structures that perform the activities. Because of this, the Chinese have no system of anatomy comparable to that of the West."

—Ted Kaptchuk, The Web That Has No Weaver

These functions are aggregated and then associated with a primary functional entity – for instance, nourishment of the tissues and maintenance of their moisture are seen as connected functions, and the entity postulated to be responsible for these functions is xuě (blood). These functional entities thus constitute *concepts* rather than something with biochemical or anatomical properties.

The primary functional entities used by traditional Chinese medicine are qì, xuě, the five zàng organs, the six fŭ organs, and the meridians which extend through the organ systems. These are all theoretically interconnected: each zàng organ is paired with a fŭ organ, which are nourished by the blood and concentrate qi for a particular function, with meridians being extensions of those functional systems throughout the body.

Concepts of the body and of disease used in TCM have notions of a pre-scientific culture, similar to European humoral theory. – TCM is characterized as full of pseudoscience. Some practitioners no longer consider yin and yang and the idea of an energy flow to apply. Scientific investigation has not found any histological or physiological evidence for traditional Chinese concepts such as *qi*, meridians, and acupuncture points. It is a generally held belief within the acupuncture community that acupuncture points and meridians structures are special conduits for electrical signals but no research has established any consistent anatomical structure or function for either acupuncture points or meridians. The scientific evidence for the anatomical existence of either meridians or acupuncture points is not compelling. Stephen Barrett of Quackwatch writes that, "TCM theory and practice are not based upon the body of knowledge related to health, disease, and health care that has been widely accepted by the scientific community. TCM practitioners disagree among themselves about how to diagnose patients and which treatments should go with which diagnoses. Even if they could agree, the TCM theories are so nebulous that no amount of scientific study will enable TCM to offer rational care."

TCM has been the subject of controversy within China. In 2006, the Chinese scholar Zhang Gongyao triggered a national debate when he published an article entitled "Farewell to Traditional Chinese Medicine," arguing that TCM was a pseudoscience that should be abolished in public healthcare and academia. The Chinese government however, interested in the opportunity of export revenues, took the stance that TCM is a science and continued to encourage its development.

Qi

TCM distinguishes many kinds of qi (pinyin: *qì*). In a general sense, qi is something that is defined by five "cardinal functions":

1. Actuation (pinyin: *tuīdòng*) – of all physical processes in the body, especially the circulation of all body fluids such as blood in their vessels. This includes actuation of the functions of the zang-fu organs and meridians.
2. Warming (pinyin: *wēnxù*) – the body, especially the limbs.
3. Defense (pinyin: *fángyù*) – against Exogenous Pathogenic Factors
4. Containment (pinyin: *gùshè*) – of body fluids, i.e., keeping blood, sweat, urine, semen, etc. from leakage or excessive emission.
5. Transformation (pinyin: *qìhuà*) – of food, drink, and breath into qi, xue (blood), and jinye ("fluids"), and/or transformation of all of the latter into each other.

Vacuity of qi will be characterized especially by pale complexion, lassitude of spirit, lack of strength, spontaneous sweating, laziness to speak, non-digestion of food, shortness of breath (especially on exertion), and a pale and enlarged tongue.

Qi is believed to be partially generated from food and drink, and partially from air (by breathing). Another considerable part of it is inherited from the parents and will be consumed in the course of life.

TCM uses special terms for qi running inside of the blood vessels and for qi that is distributed in the skin, muscles, and tissues between those. The former is called yíng-qì ; its function is to complement xuè and its nature has a strong yin aspect (although qi in general is considered to be yang). The latter is called weì-qì; its main function is defence and it has pronounced yang nature.

Qi is said to circulate in the meridians. Just as the qi held by each of the zang-fu organs, this is considered to be part of the 'principal' qi (pinyin: *yuánqì*) of the body (pinyin: *zhēn qì*, "true" qi, or pinyin: *yuán qì*, "original" qi).

Xue

In contrast to the majority of other functional entities, xuè ("blood") is correlated with a physical form – the red liquid running in the blood vessels. Its concept is, nevertheless, defined by its functions: nourishing all parts and tissues of the body, safeguarding an adequate degree of moisture, and sustaining and soothing both consciousness and sleep.

Typical symptoms of a lack of xuě (usually termed "blood vacuity" [pinyin: *xuě xū*]) are described as: Pale-white or withered-yellow complexion, dizziness, flowery vision, palpitations, insomnia, numbness of the extremities; pale tongue; "fine" pulse.

Jinye

Closely related to xuě are the jīnyė (usually translated as "body fluids"), and just like xuě they are considered to be yin in nature, and defined first and foremost by the functions of nurturing and moisturizing the different structures of the body. Their other functions are to harmonize yin and yang, and to help with the secretion of waste products.

Jīnyė are ultimately extracted from food and drink, and constitute the raw material for the production of xuě; conversely, xuě can also be transformed into jīnyė. Their palpable manifestations are all bodily fluids: tears, sputum, saliva, gastric acid, joint fluid, sweat, urine, etc.

Zang-fu

The zàng-fǔ constitute the centre piece of TCM's systematization of bodily functions. Bearing the names of organs, they are, however, only secondarily tied to (rudimentary) anatomical assumptions (the fǔ a little more, the zàng much less). As they are primarily defined by their functions, they are not equivalent to the anatomical organs–to highlight this fact, their names are usually capitalized.

The term zàng refers to the five entities considered to be yin in nature–Heart, Liver, Spleen, Lung, Kidney–, while fǔ refers to the six yang organs–Small Intestine, Large Intestine, Gallbladder, Urinary Bladder, Stomach and Sānjiaō.

The zàng's essential functions consist in production and storage of qì and xuě; they are said to regulate digestion, breathing, water metabolism, the musculoskeletal system, the skin, the sense organs, aging, emotional processes, and mental activity, among other structures and processes. The fǔ organs' main purpose is merely to transmit and digest (pinyin: *chuán-huà*) substances such as waste and food.

Since their concept was developed on the basis of Wǔ Xíng philosophy, each zàng is paired with a fǔ, and each zàng-fǔ pair is assigned to one of five elemental qualities (i.e., the Five Elements or Five Phases). These correspondences are stipulated as:

- Fire = Heart (pinyin: *xīn*) and Small Intestine (pinyin: *xiaǒcháng*) (and, secondarily, Sānjiaō ["Triple Burner"] and Pericardium [pinyin: *xīnbaò*])
- Earth = Spleen (pinyin: *pí*) and Stomach (pinyin: *weì*)
- Metal = Lung (pinyin: *feì*) and Large Intestine (pinyin: *dàcháng*)
- Water = Kidney (pinyin: *shèn*) and Bladder (pinyin: *pángguāng*)
- Wood = Liver (pinyin: *gān*) and Gallbladder (pinyin: *dān*)

The zàng-fǔ are also connected to the twelve standard meridians–each yang meridian is attached to a fǔ organ, and five of the yin meridians are attached to a zàng. As there are only five zàng but six yin meridians, the sixth is assigned to the Pericardium, a peculiar entity almost similar to the Heart zàng.

Jing-luo

Acupuncture chart from the Ming Dynasty (c. 1368–1644)

The meridians (pinyin: *jīng-luò*) are believed to be channels running from the zàng-fǔ in the interior (pinyin: *lǐ*) of the body to the limbs and joints ("the surface" [pinyin: *biaǒ*]), transporting qi and xuě. TCM identifies 12 "regular" and 8 "extraordinary" meridians; the Chinese terms being (pinyin: *shí-èr jīngmài*, lit. "the Twelve Vessels") and (pinyin: *qí jīng bā mài*) respectively. There's also a number of less customary channels branching from the "regular" meridians.

Concept of Disease

In general, disease is perceived as a disharmony (or imbalance) in the functions or interactions of yin, yang, qi, xuě, zàng-fǔ, meridians etc. and/or of the interaction between the human body and the environment. Therapy is based on which "pattern of disharmony" can be identified. Thus, "pattern discrimination" is the most important step in TCM diagnosis. It is also known to be the most difficult aspect of practicing TCM.

In order to determine which pattern is at hand, practitioners will examine things like the color and shape of the tongue, the relative strength of pulse-points, the smell of the breath, the quality of breathing or the sound of the voice. For example, depending on tongue and pulse conditions, a TCM practitioner might diagnose bleeding from the mouth and nose as: "Liver fire rushes upwards and scorches the Lung, injuring the blood vessels and giving rise to reckless pouring of blood from the mouth and nose." He might then go on to prescribe treatments designed to clear heat or supplement the Lung.

Disease Entities

In TCM, a disease has two aspects: "bìng" and "zhèng". The former is often translated as "disease entity", "disease category", "illness", or simply "diagnosis". The latter, and more important one, is usually translated as "pattern" (or sometimes also as "syndrome"). For example, the disease entity of a common cold might present with a pattern of wind-cold in one person, and with the pattern of wind-heat in another.

From a scientific point of view, most of the disease entitites (pinyin: *bìng*) listed by TCM constitute mere symptoms. Examples include headache, cough, abdominal pain, constipation etc.

Since therapy will not be chosen according to the disease entity but according to the pattern, two people with the same disease entity but different patterns will receive different therapy. Vice versa, people with similar patterns might receive similar therapy even if their disease entities are different. This is called (pinyin: *yì bìng tóng zhì, tóng bìng yì zhì,*"different diseases, same treatment; same disease, different treatments").

Patterns

In TCM, "pattern" (证, pinyin: *zhèng*) refers to a "pattern of disharmony" or "functional disturbance" within the functional entities the TCM model of the body is composed of. There are disharmony patterns of qi, xuě, the body fluids, the zàng-fŭ, and the meridians. They are ultimately defined by their symptoms and "signs" (i.e., for example, pulse and tongue findings).

In clinical practice, the identified pattern usually involves a combination of affected entities (compare with typical examples of patterns). The concrete pattern identified should account for *all* the symptoms a person has.

Six Excesses

The Six Excesses (pinyin: *lìù yín*, sometimes also translated as "Pathogenic Factors", or "Six Pernicious Influences"; pinyin: *lìù xié*, – "Six Evils" or "Six Devils") are allegorical terms used to describe disharmony patterns displaying certain typical symptoms. These symptoms resemble the effects of six climatic factors. In the allegory, these symptoms can occur because one or more of those climatic factors (called , pinyin: *lìù qì*, "the six qi") were able to invade the body surface and to proceed to the interior. This is sometimes used to draw causal relationships (i.e., prior exposure to wind/cold/etc. is identified as the cause of a disease), while other authors explicitly deny a direct cause-effect relationship between weather conditions and disease, pointing out that the Six Excesses are primarily descriptions of a certain combination of symptoms translated into a pattern of disharmony. It is undisputed, though, that the Six Excesses can manifest inside the body without an external cause. In this case, they might be denoted "internal", e.g., "internal wind" or "internal fire (or heat)".

The Six Excesses and their characteristic clinical signs are:

1. Wind (pinyin: *fēng*): rapid onset of symptoms, wandering location of symptoms, itching, nasal congestion, "floating" pulse; tremor, paralysis, convulsion.
2. Cold (pinyin: *hán*): cold sensations, aversion to cold, relief of symptoms by warmth, watery/clear excreta, severe pain, abdominal pain, contracture/hypertonicity of muscles, (slimy) white tongue fur, "deep"/"hidden" or "string-like" pulse, or slow pulse.
3. Fire/Heat (pinyin: *huǒ*): aversion to heat, high fever, thirst, concentrated urine, red face, red tongue, yellow tongue fur, rapid pulse. (Fire and heat are basically seen to be the same)
4. Dampness (pinyin: *shī*): sensation of heaviness, sensation of fullness, symptoms of Spleen dysfunction, greasy tongue fur, "slippery" pulse.
5. Dryness (pinyin: *zào*): dry cough, dry mouth, dry throat, dry lips, nosebleeds, dry skin, dry stools.
6. Summerheat (pinyin: *shǔ*): either heat or mixed damp-heat symptoms.

Six-Excesses-patterns can consist of only one or a combination of Excesses (e.g., wind-cold, wind-damp-heat). They can also transform from one into another.

Typical Examples of Patterns

For each of the functional entities (qi, xuě, zàng-fǔ, meridians etc.), typical disharmony patterns are recognized; for example: qi vacuity and qi stagnation in the case of qi; blood vacuity, blood stasis, and blood heat in the case of xuě; Spleen qi vacuity, Spleen yang vacuity, Spleen qi vacuity with down-bearing qi, Spleen qi vacuity with lack of blood containment, cold-damp invasion of the Spleen, damp-heat invasion of Spleen and Stomach in case of the Spleen zàng; wind/cold/damp invasion in the case of the meridians.

TCM gives detailed prescriptions of these patterns regarding their typical symptoms, mostly including characteristic tongue and/or pulse findings. For example:

- "Upflaming Liver fire" (pinyin: *gānhuǒ shàng yán*): Headache, red face, reddened eyes, dry mouth, nosebleeds, constipation, dry or hard stools, profuse menstruation, sudden tinnitus or deafness, vomiting of sour or bitter fluids, expectoration of blood, irascibility, impatience; red tongue with dry yellow fur; slippery and string-like pulse.

Eight Principles of Diagnosis

The process of determining which actual pattern is on hand is called (pinyin: *biàn zhèng*, usually translated as "pattern diagnosis", "pattern identification" or "pattern discrimination"). Generally, the first and most important step in pattern diagnosis is an evaluation of the present signs and symptoms on the basis of the "Eight Principles" (pinyin: *bā gāng*). These eight principles refer to four pairs of fundamental qualities of a disease: exterior/interior, heat/cold, vacuity/repletion, and yin/yang. Out of these, heat/cold and vacuity/repletion have the biggest clinical importance. The yin/yang quality, on the other side, has the smallest importance and is somewhat seen aside from the other three pairs, since it merely presents a general and vague conclusion regarding what other qualities are found. In detail, the Eight Principles refer to the following:

- *Yin and yang* are universal aspects all things can be classified under, this includes diseases in general as well as the Eight Principles' first three couples. For example, cold is identified to be a yin aspect, while heat is attributed to yang. Since descriptions of patterns in terms of yin and yang lack complexity and clinical practicality, though, patterns are usually not labelled this way anymore. Exceptions are vacuity-cold and repletion-heat patterns, who are sometimes referred to as "yin patterns" and "yang patterns" respectively.

- *Exterior* (pinyin: *biǎo*) refers to a disease manifesting in the superficial layers of the body – skin, hair, flesh, and meridians. It is characterized by aversion to cold and/or wind, headache, muscle ache, mild fever, a "floating" pulse, and a normal tongue appearance.

- *Interior* (pinyin: *lǐ*) refers to disease manifestation in the zàng-fǔ, or (in a wider sense) to any disease that can not be counted as exterior. There are no generalized characteristic symptoms of interior patterns, since they'll be determined by the affected zàng or fǔ entity.

- *Cold* (pinyin: *hán*) is generally characterized by aversion to cold, absence of thirst, and a white tongue fur. More detailed characterization depends on whether cold is coupled with vacuity or repletion.

- *Heat* (pinyin: *rè*) is characterized by absence of aversion to cold, a red and painful throat, a dry tongue fur and a rapid and floating pulse, if it falls together with an exterior pattern. In all other cases, symptoms depend on whether heat is coupled with vacuity or repletion.

- *Deficiency* (pinyin: *xū*), can be further differentiated into deficiency of qi, xuě, yin and yang, with all their respective characteristic symptoms. Yin deficiency can also cause "empty-heat".

- *Excess* (pinyin: *shí*) generally refers to any disease that can't be identified as a deficient pattern, and usually indicates the presence of one of the Six Excesses, or a pattern of stagnation (of qi, xuě, etc.). In a concurrent exterior pattern, excess is characterized by the absence of sweating.

After the fundamental nature of a disease in terms of the Eight Principles is determined, the investigation focuses on more specific aspects. By evaluating the present signs and symptoms against the background of typical disharmony patterns of the various entities, evidence is collected whether or how specific entities are affected. This evaluation can be done

1. in respect of the meridians (pinyin: *jīng-luò biàn zhèng*)
2. in respect of qi (pinyin: *qì xuě biàn zhèng*)
3. in respect of xuě (pinyin: *qì xuě biàn zhèng*)
4. in respect of the body fluids (pinyin: *jīn-yė biàn zhèng*)
5. in respect of the zàng-fǔ (pinyin: *zàng-fǔ biàn zhèng*) – very similar to this, though less specific, is disharmony pattern description in terms of the Five Elements [pinyin: *wǔ xíng biàn zhèng*])

There are also three special pattern diagnosis systems used in case of febrile and infectious diseases only ("Six Channel system" or "six division pattern" [pinyin: *liù jīng biàn zhèng*]; "Wei Qi Ying Xue system" or "four division pattern" [pinyin: *weì qì yíng xuě biàn zhèng*]; "San Jiao system" or "three burners pattern" [pinyin: *sānjiaō biàn zhèng*]).

Considerations of Disease Causes

Although TCM and its concept of disease do not strongly differentiate between cause and effect, pattern discrimination can include considerations regarding the disease cause; this is called (pinyin: *bìngyīn biàn zhèng*, "disease-cause pattern discrimination").

There are three fundamental categories of disease causes (pinyin: *sān yīn*) recognized:

1. external causes: these include the Six Excesses and "Pestilential Qi".
2. internal causes: the "Seven Affects" (pinyin: *qì qíng*, sometimes also translated as "Seven Emotions") – joy, anger, brooding, sorrow, fear, fright and grief. These are believed to be able to cause damage to the functions of the zàng-fú, especially of the Liver.
3. non-external-non-internal causes: dietary irregularities (especially: too much raw, cold, spicy, fatty or sweet food; voracious eating; too much alcohol), fatigue, sexual intemperance, trauma, and parasites (pinyin: *chóng*).

Diagnostics

In TCM, there are five diagnostic methods: inspection, auscultation, olfaction, inquiry, and palpation.

- Inspection focuses on the face and particularly on the tongue, including analysis of the tongue size, shape, tension, color and coating, and the absence or presence of teeth marks around the edge.
- Auscultation refers to listening for particular sounds (such as wheezing).
- Olfaction refers to attending to body odor.
- Inquiry focuses on the "seven inquiries", which involve asking the person about the regularity, severity, or other characteristics of: chills, fever, perspiration, appetite, thirst, taste, defecation, urination, pain, sleep, menses, leukorrhea.
- Palpation which includes feeling the body for tender A-shi points, and the palpation of the wrist pulses as well as various other pulses, and palpation of the abdomen.

Tongue and Pulse

Examination of the tongue and the pulse are among the principal diagnostic methods in TCM. Certain sectors of the tongue's surface are believed to correspond to the zàng-fŭ. For example, teeth marks on one part of the tongue might indicate a problem with the Heart, while teeth marks on another part of the tongue might indicate a problem with the Liver.

Pulse palpation involves measuring the pulse both at a superficial and at a deep level at three different locations on the radial artery (*Cun, Guan, Chi,* located two fingerbreadths from the wrist crease, one fingerbreadth from the wrist crease, and right at the wrist crease, respectively, usually palpated with the index, middle and ring finger) of each arm, for a total of twelve pulses, all of which are thought to correspond with certain zàng-fǔ. The pulse is examined for several characteristics including rhythm, strength and volume, and described with qualities like "floating, slippery, bolstering-like, feeble, thready and quick"; each of these qualities indicate certain disease patterns. Learning TCM pulse diagnosis can take several years.

Herbal Medicine

The term "herbal medicine" is somewhat misleading in that, while plant elements are by far the most commonly used substances in TCM, other, non-botanic substances are used as well: animal, human, and mineral products are also utilized. Thus, the term "medicinal" (instead of herb) is usually preferred.

Assorted dried plant and animal parts used in traditional Chinese medicines, clockwise from top left corner: dried Lingzhi (lit. "spirit mushrooms"), ginseng, Luo Han Guo, turtle shell underbelly (plastron), and dried curled snakes.

A bile bear in a "crush cage" on Huizhou Farm, China.

Prescriptions

Typically, one batch of medicinals is prepared as a decoction of about 9 to 18 substances. Some of these are considered as main herbs, some as ancillary herbs; within the ancillary herbs, up to three categories can be distinguished.

Raw Materials

There are roughly 13,000 medicinals used in China and over 100,000 medicinal recipes recorded in the ancient literature. Plant elements and extracts are by far the most common elements used.

In the classic *Handbook of Traditional Drugs* from 1941, 517 drugs were listed – out of these, 45 were animal parts, and 30 were minerals.

Animal Substances

Some animal parts used as medicinals can be considered rather strange such as cows' gallstones, hornet's nest, leech, and scorpion. Other examples of animal parts include horn of the antelope or buffalo, deer antlers, testicles and penis bone of the dog, and snake bile. Some TCM textbooks still recommend preparations containing animal tissues, but there has been little research to justify the claimed clinical efficacy of many TCM animal products.

Some medicinals can include the parts of endangered species, including tiger bones and rhinoceros horn which is used for many ailments (though not as an aphrodisiac as is commonly misunderstood by the West). The black market in rhinoceros horn (driven not just by TCM but also unrelated status-seeking) has reduced the world's rhino population by more than 90 percent over the past 40 years. Concerns have also arisen over the use of pangolin scales, turtle plastron, seahorses, and the gill plates of mobula and manta rays. Poachers hunt restricted or endangered species animals to supply the black market with TCM products. There is no scientific evidence of efficacy for tiger medicines. Concern over China considering to legalize the trade in tiger parts prompted the 171-nation Convention on International Trade in Endangered Species (CITES) to endorse a decision opposing the resurgence of trade in tigers. Fewer than 30,000 saiga antelopes remain, which are exported to China for use in traditional fever therapies. Organized gangs illegally export the horn of the antelopes to China. The pressures on seahorses (Hippocampus spp.) used in traditional medicine is large; tens of millions of animals are unsustainably caught annually. Many species of syngnathid are currently part of the IUCN Red List of Threatened Species or national equivalents.

Since TCM recognizes bear bile as a medicinal, more than 12,000 asiatic black bears are held in bear farms. The bile is extracted through a permanent hole in the abdomen leading to the gall bladder, which can cause severe pain. This can lead to bears trying to kill themselves. As of 2012, approximately 10,000 bears are farmed in China for their bile. This practice has spurred public outcry across the country. The bile is collected from live bears via a surgical procedure. The deer penis is believed to have therapeutic benefits according to traditional Chinese medicine. It is typically very big and, proponents believe, in order to preserve its properties, it should be extracted from a living deer. Medicinal tiger parts from poached animals include tiger penis, believed to improve virility, and tiger eyes. The illegal trade for tiger parts in China has driven the species to near-extinction because of its popularity in traditional medicine. Laws protecting even critically endangered species such as the Sumatran tiger fail to stop the display and sale of these items in open markets. Shark fin soup is traditionally regarded in Chinese medicine as beneficial for health in East Asia, and its status as an elite dish has led to huge demand with the increase of affluence in China, devastating shark populations. The shark fins have been a part of traditional Chinese medicine for centuries. Shark finning is banned in many countries, but the trade is thriving in Hong Kong and China, where the fins are part of shark fin soup, a dish considered a delicacy, and used in some types of traditional Chinese medicine.

The tortoise (Freshwater turtle - guiban) and the turtle (Chinese softshell turtle - biejia) species used in traditional Chinese medicine are raised on farms, while restrictions are made on the accumulation and export of other endangered species. However, issues concerning the overexploitation of Asian turtles in China have not been completely solved. Australian scientists have developed methods to identify medicines containing DNA traces of endangered species. Finally, although not an endangered species, sharp rises in exports of donkeys and donkey hide from Africa to China to make the traditional remedy ejiao have prompted export restrictions by some African countries.

Human Body Parts

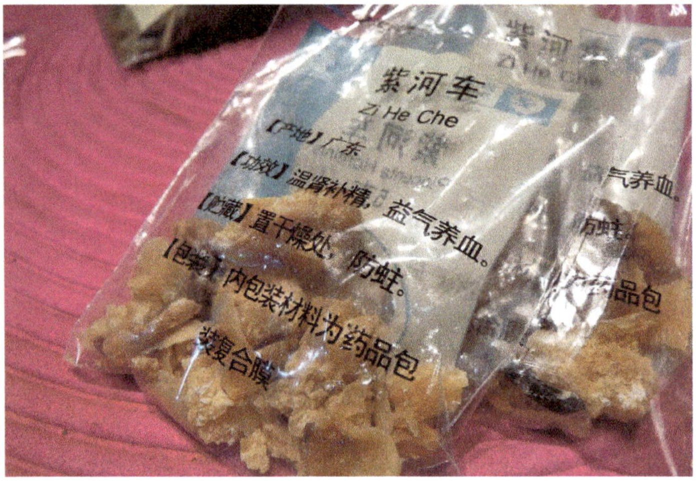

Dried human placenta (Ziheche (车) is used in traditional Chinese medicine.

Traditional Chinese Medicine also includes some human parts: the classic Materia medica (Bencao Gangmu) describes the use of 35 human body parts and excreta in medicines, including bones, fingernail, hairs, dandruff, earwax, impurities on the teeth, feces, urine, sweat, organs, but most are no longer in use.

Human placenta has been used an ingredient in certain traditional Chinese medicines, including using dried human placenta, known as "Ziheche", to treat infertility, impotence and other conditions. The consumption of the human placenta is a potential source of infection.

Traditional Categorization

The traditional categorizations and classifications that can still be found today are:

The classification according to the Four Natures (pinyin: *sì qì*): hot, warm, cool, or cold (or, neutral in terms of temperature) and hot and warm herbs are used to treat cold diseases, while cool and cold herbs are used to treat heat diseases.

The classification according to the Five Flavors, (pinyin: *wǔ wèi*, sometimes also translated as Five Tastes): acrid, sweet, bitter, sour, and salty. Substances may also have more than one flavor, or none (i.e., a "bland" flavor). Each of the Five Flavors corresponds to one of zàng organs, which in turn corresponds to one of the Five Phases. A flavor implies certain properties and therapeutic actions of a substance; e.g., saltiness drains downward and softens hard masses, while sweetness is supplementing, harmonizing, and moistening.

The classification according to the meridian – more precise, the zàng-organ including its associated meridian – which can be expected to be primarily affected by a given medicinal.

The categorization according to the specific function mainly include: exterior-releasing or exterior-resolving, heat-clearing, downward-draining, or precipitating wind-damp-dispelling, dampness-transforming, promoting the movement of water and percolating dampness or dampness-percolating, interior-warming, qi-regulating or qi-rectifying, dispersing food accumulation or food-dispersing, worm-expelling, stopping bleeding or blood-stanching, quickening the Blood and dispelling stasis or blood-quickening, transforming phlegm, stopping coughing and calming wheezing or phlegm-transforming and cough- and panting-suppressing, Spirit-quieting, calming the liver and expelling wind or liver-calming and wind-extinguishingl orifice-openingl supplementing which includes qi-supplementing, blood-nourishing, yin-enriching, and yang-fortifying, astriction-promoting or securing and astringing, vomiting-inducing, and substances for external application.

Efficacy

As of 2007 there were not enough good-quality trials of herbal therapies to allow their effectiveness to be determined. A high percentage of relevant studies on traditional Chinese medicine are in Chinese databases. Fifty percent of systematic reviews on TCM did not search Chinese databases, which could lead to a bias in the results. Many systematic reviews of TCM interventions published in Chinese journals are incomplete, some contained errors or were misleading. The herbs recommended by traditional Chinese practitioners in the US are unregulated.

A 2013 review found the data too weak to support use of Chinese herbal medicine (CHM) for benign prostatic hyperplasia. A 2013 review found the research on the benefit and safety of CHM for idiopathic sudden sensorineural hearing loss is of poor quality and cannot be relied upon to support their use. A 2013 Cochrane review found inconclusive evidence that CHM reduces the severity of eczema. The traditional medicine ginger, which has shown anti-inflammatory properties in laboratory experiments, has been used to treat rheumatism, headache and digestive and respiratory issues, though there is no firm evidence supporting these uses. A 2012 Cochrane review found no difference in decreased mortality when Chinese herbs were used alongside Western medicine versus Western medicine exclusively. A 2012 Cochrane review found insufficient evidence to support the use of TCM for people with adhesive small bowel obstruction. A 2011 review found low quality evidence that suggests CHM improves the symptoms of Sjogren's syndrome. A 2010 review found TCM seems to be effective for the treatment of fibromyalgia but the findings were of insufficient methodological rigor. A 2009 Cochrane review found insufficient evidence to recommend the use of TCM for the treatment of epilepsy. A 2008 Cochrane review found promising evidence for the use of Chinese herbal medicine in relieving painful menstruation, but the trials assessed were of such low methodological quality that no conclusion could be drawn about the remedies' suitability as a recommendable treatment option. Turmeric has been used in traditional Chinese medicine for centuries to treat various conditions. This includes jaundice and hepatic disorders, rheumatism, anorexia, diabetic wounds, and menstrual complications. Most of its effects have been attributed to curcumin. Research that curcumin shows strong anti-inflammatory and antioxidant activities have instigated mechanism of action studies on the possibility for cancer and inflammatory diseases prevention and treatment. It also exhibits immunomodulatory effects. A 2005 Cochrane review found insufficient evidence for the use of CHM in HIV-infected people and people with AIDS.

Drug Research

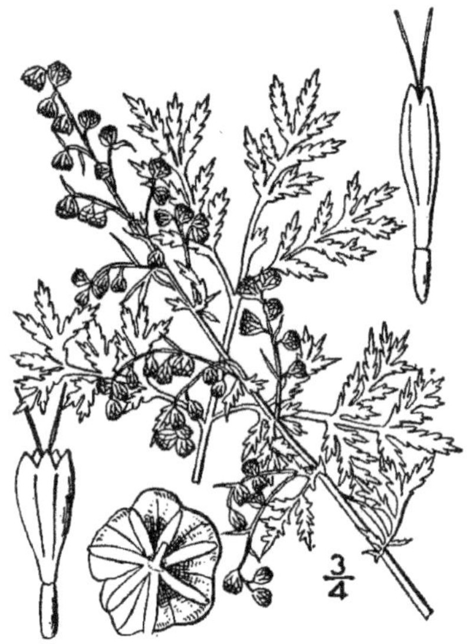

Artemisia annua is traditionally used to treat fever.
It has been found to have antimalarial properties.

With an eye to the enormous Chinese market, pharmaceutical companies have explored the potential for creating new drugs from traditional remedies. A *Nature* editorial described TCM as "fraught with pseudoscience", and stated that having "no rational mechanism of action for most of its therapies" is the "most obvious answer" to why its study didn't provide a "flood of cures", while advocates responded that "researchers are missing aspects of the art, notably the interactions between different ingredients in traditional therapies."

One of the few successes was the development in the 1970s of the antimalarial drug artemisinin, which is a processed extract of *Artemisia annua*, a herb traditionally used as a fever treatment. Researcher Tu Youyou discovered that a low-temperature extraction process could isolate an effective antimalarial substance from the plant. She says she was influenced by a traditional source saying that this herb should be steeped in cold water, after initially finding high-temperature extraction unsatisfactory. The extracted substance, once subject to detoxification and purification processes, is a usable antimalarial drug – a 2012 review found that artemisinin-based remedies were the most effective drugs for the treatment of malaria. For her work on malaria, Tu received the 2015 Nobel Prize in Physiology or Medicine. Despite global efforts in combating malaria, it remains a large burden for the population. Although WHO recommends artemisinin-based remedies for treating uncomplicated malaria, artemisinin resistance can no longer be ignored.

Also in the 1970s Chinese researcher Zhang TingDong and colleagues investigated the potential use of the traditionally used substance arsenic trioxide to treat acute promyelocytic leukemia (APL). Building on his work, research both in China and the West eventually led to the development of the drug Trisenox, which was approved for leukemia treatment by the FDA in 2000.

Huperzine A, which is extracted from traditional herb *Huperzia serrata*, has attracted the interest of medical science because of alleged neuroprotective properties. Despite earlier promising results, a 2013 systematic review and meta-analysis found "Huperzine A appears to have beneficial effects on improvement of cognitive function, daily living activity, and global clinical assessment in participants with Alzheimer's disease. However, the findings should be interpreted with caution due to the poor methodological quality of the included trials."

Ephedrine in its natural form, known as *má huáng* in traditional Chinese medicine, has been documented in China since the Han dynasty (206 BC – 220 AD) as an antiasthmatic and stimulant. In 1885, the chemical synthesis of ephedrine was first accomplished by Japanese organic chemist Nagai Nagayoshi based on his research on Japanese and Chinese traditional herbal medicines

Cost-effectiveness

A 2012 systematic review found there is a lack of available cost-effectiveness evidence in TCM.

Safety

Galena (lead ore) is part of TCM.

From the earliest records regarding the use of medicinals to today, the toxicity of certain substances has been described in all Chinese materiae medicae. Since TCM has become more popular in the Western world, there are increasing concerns about the potential toxicity of many traditional Chinese medicinals including plants, animal parts and minerals. Traditional Chinese herbal remedies are conveniently available from grocery stores in most Chinese neighborhoods; some of these items may contain toxic ingredients, are imported into the U.S. illegally, and are associated with claims of therapeutic benefit without evidence. For most medicinals, efficacy and toxicity testing are based on traditional knowledge rather than laboratory analysis. The toxicity in some cases could be confirmed by modern research (i.e., in scorpion); in some cases it couldn't (i.e., in *Curculigo*). Traditional herbal medicines can contain extremely toxic chemicals and heavy metals, and naturally occurring toxins, which can cause illness, exacerbate pre-existing poor health or result in death. Botanical misidentification of plants can cause toxic reactions in humans. The description on some plants used in traditional Chinese medicine have changed, leading to unintended intoxication of the wrong plants. A concern is also contaminated herbal medicines with microorganisms and fungal toxins, including aflatoxin. Traditional herbal medicines are sometimes contaminated with toxic heavy metals, including lead, arsenic, mercury and cadmium, which inflict serious health risks to consumers. Also, adulteration of some herbal medicine preparations with conventional drugs which may cause serious adverse effects, such as corticosteroids, phenylbutazone, phenytoin, and glibenclamide, has been reported.

Substances known to be potentially dangerous include *Aconitum*, secretions from the Asiatic toad, powdered centipede, the Chinese beetle (*Mylabris phalerata*), certain fungi, *Aristolochia*, *Aconitum*, Arsenic sulfide (Realgar), mercury sulfide, and cinnabar. Asbestos ore (Actinolite, Yang Qi Shi) is used to treat impotence in TCM. Due to galena's (litharge, lead(II) oxide) high lead content, it is known to be toxic. Lead, mercury, arsenic, copper, cadmium, and thallium have been detected in TCM products sold in the U.S. and China.

To avoid its toxic adverse effects *Xanthium sibiricum* must be processed. Hepatotoxicity has been reported with products containing *Polygonum multiflorum*, glycyrrhizin, *Senecio* and *Symphytum*. The herbs indicated as being hepatotoxic included *Dictamnus dasycarpus*, *Astragalus membranaceous*, and *Paeonia lactiflora*. Contrary to popular belief, *Ganoderma lucidum* mushroom extract, as an adjuvant for cancer immunotherapy, appears to have the potential for toxicity. A 2013 review suggested that although the antimalarial herb *Artemisia annua* may not cause hepatotoxicity, haematotoxicity, or hyperlipidemia, it should be used cautiously during pregnancy due to a potential risk of embryotoxicity at a high dose.

However, many adverse reactions are due to misuse or abuse of Chinese medicine. For example, the misuse of the dietary supplement *Ephedra* (containing ephedrine) can lead to adverse events including gastrointestinal problems as well as sudden death from cardiomyopathy. Products adulterated with pharmaceuticals for weight loss or erectile dysfunction are one of the main concerns. Chinese herbal medicine has been a major cause of acute liver failure in China.

Acupuncture and Moxibustion

Acupuncture is the insertion of needles into superficial structures of the body (skin, subcutaneous tissue, muscles) – usually at acupuncture points (acupoints) – and their subsequent manipulation; this aims at influencing the flow of qi. According to TCM it relieves pain and treats (and prevents) various diseases.

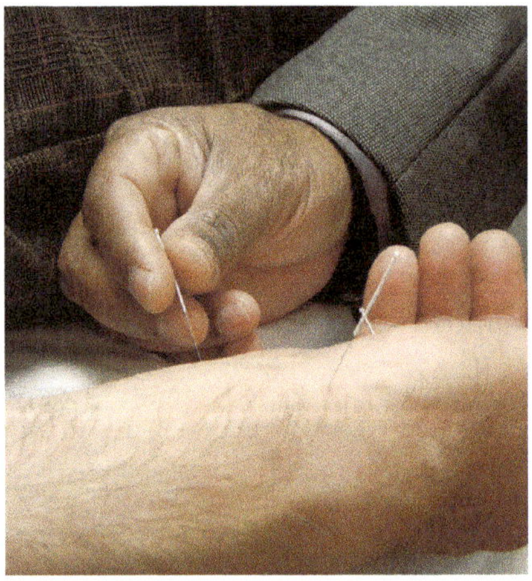

Needles being inserted into the skin.

Acupuncture is often accompanied by moxibustion – (pinyin: *zhēnjiǔ*) literally meaning "acupuncture-moxibustion" – which involves burning mugwort on or near the skin at an acupuncture point. According to the American Cancer Society, "available scientific evidence does not support claims that moxibustion is effective in preventing or treating cancer or any other disease".

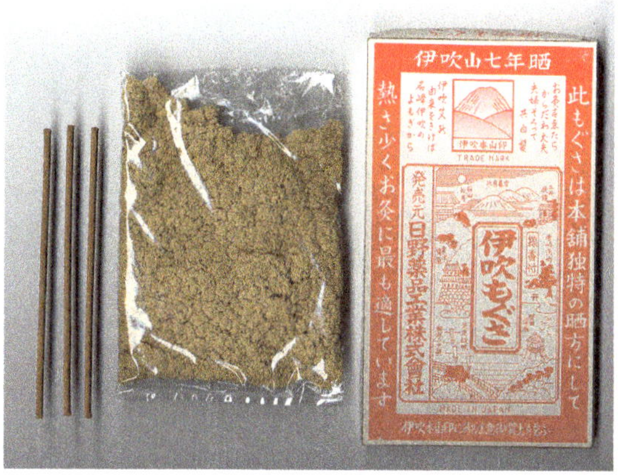

Traditional moxibustion set from Ibuki (Japan)

In electroacupuncture, an electric current is applied to the needles once they are inserted, in order to further stimulate the respective acupuncture points.

Efficacy

A 2013 editorial by Steven P. Novella and David Colquhoun found that the inconsistency of results of acupuncture studies (i.e. acupuncture relieved pain in some conditions but had no effect in other very similar conditions) suggests false positive results, which may be caused by factors like biased study designs, poor blinding, and the classification of electrified needles (a type of TENS) as a form of acupuncture. The same editorial suggested that given the inability to find consistent results despite more than 3,000 studies of acupuncture, the treatment seems to be a placebo effect and the existing equivocal positive results are noise one expects to see after a large number of studies are performed on an inert therapy. The editorial concluded that the best controlled studies showed a clear pattern, in which the outcome does not rely upon needle location or even needle insertion, and since "these variables are those that define acupuncture, the only sensible conclusion is that acupuncture does not work."

A 2012 meta-analysis concluded that the mechanisms of acupuncture "are clinically relevant, but that an important part of these total effects is not due to issues considered to be crucial by most acupuncturists, such as the correct location of points and depth of needling ... [but are] ... associated with more potent placebo or context effects". Commenting on this meta-analysis, both Edzard Ernst and David Colquhoun said the results were of negligible clinical significance.

A 2011 overview of Cochrane reviews found high quality evidence that suggests acupuncture is effective for some but not all kinds of pain. A 2010 systematic review found that there is evidence "that acupuncture provides a short-term clinically relevant effect when compared with a waiting list control or when acupuncture is added to another intervention" in the treatment of chronic low back pain. Two review articles discussing the effectiveness of acupuncture, from 2008 and 2009, have concluded that there is not enough evidence to conclude that it is effective beyond the placebo effect.

Acupuncture is generally safe when administered using Clean Needle Technique (CNT). Although serious adverse effects are rare, acupuncture is not without risk. Severe adverse effects, including death, have continued to be reported.

Tui Na

An example of a Traditional Chinese medicine used in Tui Na

Tui na is a form of massage akin to acupressure (from which shiatsu evolved). Asian massage is typically administered with the person fully clothed, without the application of grease or oils. Techniques employed may include thumb presses, rubbing, percussion, and assisted stretching.

Qigong

Qìgōng is a TCM system of exercise and meditation that combines regulated breathing, slow movement, and focused awareness, purportedly to cultivate and balance qi. One branch of qigong is qigong massage, in which the practitioner combines massage techniques with awareness of the acupuncture channels and points.

Other Therapies

Cupping

Cupping (pinyin: báguàn) is a type of Chinese massage, consisting of placing several glass "cups" (open spheres) on the body. A match is lit and placed inside the cup and then removed before placing the cup against the skin. As the air in the cup is heated, it expands, and after placing in the skin, cools, creating lower pressure inside the cup that allows the cup to stick to the skin via suction. When combined with massage oil, the cups can be slid around the back, offering "reverse-pressure massage".

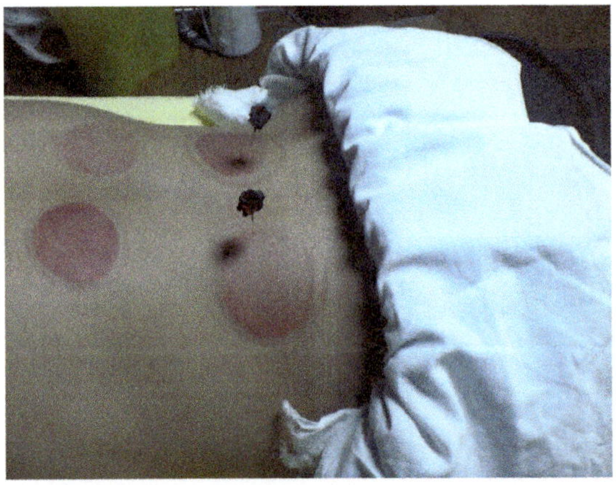
Acupuncture and moxibustion after cupping in Japan

It has not been found to be effective for the treatment of any disease. The 2008 *Trick or Treatment* book said that no evidence exists of any beneficial effects of cupping for any medical condition.

Gua Sha

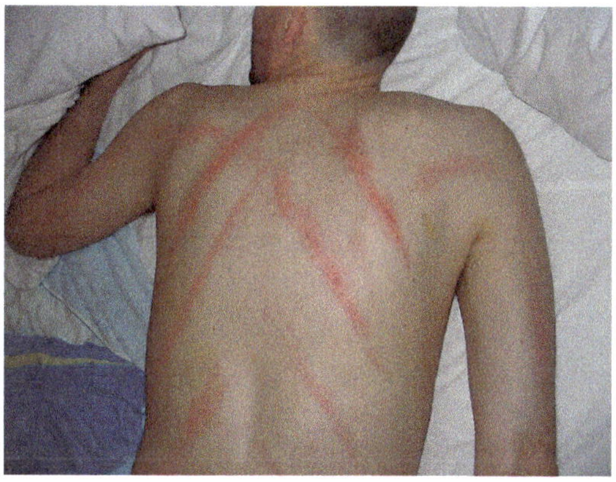

Gua Sha

Gua Sha (Chinese: pinyin: guāshā) is abrading the skin with pieces of smooth jade, bone, animal tusks or horns or smooth stones; until red spots then bruising cover the area to which it is done. It is believed that this treatment is for almost any ailment including cholera. The red spots and bruising take 3 to 10 days to heal, there is often some soreness in the area that has been treated.

Die-da

Diē-dá or bone-setting is usually practiced by martial artists who know aspects of Chinese medicine that apply to the treatment of trauma and injuries such as bone fractures, sprains, and bruises. Some of these specialists may also use or recommend other disciplines of Chinese medical therapies (or Western medicine in modern times) if serious injury is involved. Such practice of bone-setting is not common in the West.

Chinese Food Therapy

Traditional Chinese characters and for the words *yin* and *yang* denote different classes of foods, and it is important to consume them in a balanced fashion. The meal sequence should also observe these classes:

> In the Orient, it is traditional to eat yang before yin. Miso soup (yang – fermented soybean protein) for breakfast; raw fish (more yang protein); and then the vegetables which are yin.

Regulations

Many governments have enacted laws to regulate TCM practice.

Australia

From 1 July 2012 Chinese medicine practitioners must be registered under the national registration and accreditation scheme with the Chinese Medicine Board of Australia and meet the Board's Registration Standards, in order to practice in Australia.

Canada

TCM is regulated in five provinces in Canada: Alberta, British Columbia, Ontario, Quebec, and Newfoundland.

Hong Kong

The Chinese Medicine Council of Hong Kong was established in 1999. It regulates the medicinals and professional standards for TCM practitioners. All TCM practitioners in Hong Kong are required to register with the Council. The eligibility for registration includes a recognised 5-year university degree of TCM, a 30-week minimum supervised clinical internship, and passing the licensing exam.

Malaysia

The Traditional and Complementary Medicine Bill was passed by Parliament in 2012 establishing the Traditional and Complementary Medicine Council to register and regulate traditional and complementary medicine practitioners, including traditional Chinese medicine practitioners as well as other traditional and complementary medicine practitioners such as those in traditional Malay medicine and traditional Indian medicine.

Singapore

The TCM Practitioners Act was passed by Parliament in 2000 and the TCM Practitioners Board was established in 2001 as a statutory board under the Ministry of Health, to register and regulate TCM practitioners. The requirements for registration include possession of a diploma or degree from a TCM educational institution/university on a gazetted list, either structured TCM clinical training at an approved local TCM educational institution or foreign TCM registration together with supervised TCM clinical attachment/practice at an approved local TCM clinic, and upon meeting these requirements, passing the Singapore TCM Physicians Registration Examination (STRE) conducted by the TCM Practitioners Board.

United States

As of July 2012, only six states do not have existing legislation to regulate the professional practice of TCM. These six states are Alabama, Kansas, North Dakota, South Dakota, Oklahoma, and Wyoming. In 1976, California established an Acupuncture Board and became the first state licensing professional acupuncturists.

Indonesia

All traditional medicines, including TCM, are regulated on Indonesian Minister of Health Regulation in 2013 about Traditional Medicine. Traditional Medicine License (*Surat Izin Pengobatan Tradisional* -SIPT) will be granted to the practitioners whose methods are scientifically recognized as safe and bring the benefit for health. The TCM clinics are registered but there is no explicit regulation for it. The only TCM method which is accepted by medical logic and is empirically proofed is acupuncture. The acupuncturists can get SIPT and participate on health care facilities.

The Chinese traditional medicine at one of Chinese traditional medicine shop at Jagalan Road, Surabaya, Indonesia.

Angelica Sinensis

Angelica sinensis, commonly known as *dong quai* or "female ginseng" is a herb from the family Apiaceae, indigenous to China. *Angelica sinensis* grows in cool high altitude mountains in China, Japan, and Korea. The yellowish brown root of the plant is harvested in fall and is a well-known Chinese medicine used over thousands of years.

Pharmacology

Use in Traditional Chinese Medicine

The dried root of *A. sinensis* is commonly known as Chinese angelica (pinyin: *dānguī*; Pe̍h-ōe-jī: *tong-kui*) and is widely used in Chinese traditional medicine for women's health, cardiovascular conditions, osteoarthrosis, inflammation, headache, infections, mild anemia, fatigue and high blood pressure. The dong quai means that a husband shall return back to his wife, which is implicitly said to help women's sexual health.

Dong quai is used for menopause vasomotor symptoms such as hot flashes. However, a randomized, double blind, placebo-controlled clinical trial showed that *dong quai* was no more effective than placebo.

Potential anti-osteoporotic effects of *dong quai* independent of any estrogen mechanism were evaluated in rat models which showed that the extract of *A. sinensis* may prevent the bone loss. However, more high quality human evidence is needed to confirm same anti-osteoporotic effects of *dong quai* in humans.

Dong quai contains a chemical compound called butylidenephthalide which has antispasmodic activity *in vitro* and might relieve dysmenorrhoea muscle cramps by relaxing the uterus muscle. However, this claim lacks evidence of effectiveness in human clinical trials.

In an animal study, *Dong quai* also induced hair growth via the Inhibition of Apoptosis Signaling.

Overall, the U.S. National Library of Medicine states that more evidence is needed to rate the effectiveness of *dong quai* for most uses.

Adverse Effects

There is evidence that *A. sinensis* may affect the muscles of the uterus. Women who are pregnant or planning on becoming pregnant should not use *A. sinensis*, because it may induce a miscarriage. Taking *A. sinensis* can cause skin to become extra sensitive to the sun, leading to a greater risk for skin cancer. One case of gynaecomastia has been reported following consumption of *dong quai* root powder pills. Large and prolonged doses of the plant is not advised as it contains compounds that are considered carcinogenic.

Drug Interactions

A. sinensis may increase the anticoagulant effects of the drug warfarin (as it contains coumarins) and consequently increase the risk of bleeding.

Due to the antiplatelet and anticoagulant effects of *A. sinensis*, it should be taken with caution with herbs or supplements (such as ginkgo, garlic, and ginger) that may slow blood clotting to reduce the possible risk of bleeding and bruising.

Chemistry

The plant's chemical constituents include phytosterols, polysaccharides, ligustilit, b-butyl phtalit, cnidilit, isoenidilit, p-cymen, ferulate, and flavonoids. When isolated from the plant, one of the chemicals, angelica polysaccharide sulfate, has *in vitro* antioxidant activity.

Goji

Ripe goji berries on the vine

Goji, goji berry, or wolfberry (pinyin: *gǒuqǐ*) is the fruit of either the *Lycium barbarum* or *Lycium chinense*, two closely related species of boxthorn in the nightshade family, Solanaceae. The family also includes the potato, tomato, eggplant, belladonna, chili pepper, and tobacco. The two species are native to Asia.

Dried goji berries

Description

Wolfberry species are deciduous woody perennial plants, growing 1–3 m high. *L. chinense* is grown in the south of China and tends to be somewhat shorter, while *L. barbarum* is grown in the north, primarily in the Ningxia Hui Autonomous Region, and tends to be somewhat taller.

Leaves and Flowers

L. chinense leaves and flower

Wolfberry leaves form on the shoot either in an alternating arrangement or in bundles of up to three, each having either a lanceolate (shaped like a spearhead longer than it is wide) or ovate (egg-like) shape. Leaf dimensions are 7.0 cm long by 3.5 cm broad with blunted or rounded tips.

The flowers grow in groups of one to three in the leaf axils. The calyx (eventually ruptured by the growing berry) consists of bell-shaped or tubular sepals forming short, triangular lobes. The corollae are lavender or light purple, 9–14 mm wide with five or six lobes shorter than the tube. The stamens are structured with filaments longer than the anthers. The anthers are longitudinally dehiscent.

In the Northern Hemisphere, flowering occurs from June through September and berry maturation from August to October, depending on the latitude, altitude, and climate.

L. barbarum leaves and flower

Fruit

These species produce a bright orange-red, ellipsoid berry 1–2 cm in diameter. The number of seeds in each berry varies widely based on cultivar and fruit size, containing 10–60 tiny yellow seeds that are compressed with a curved embryo. The berries ripen from July to October in the Northern Hemisphere.

Naming

Lycium, the genus name, is derived from the ancient southern Anatolian region of Lycia. The fruit is known in pharmacological references as *Lycii fructus*, which is Latin for "*Lycium* fruit".

"Wolfberry", a commonly used English name, has unknown origin, perhaps confusion over the genus name, *Lycium*, which resembles *lycos*, the Greek word for wolf.

In the English-speaking world, the name "goji berry" has been used starting around 2000. The word "goji" is an approximation of the pronunciation of *gǒuqǐ*, the name for *L. chinense* in several Chinese dialects, including Hokkien and Shanghainese.

In the official classification, *L. barbarum* is known as matrimony vine and *L. chinese* as Chinese desert-thorn.

Significance

Since the early 21st century, interest has increased in wolfberries for their novelty and supposed nutrient value. They have been termed a superfruit, which has led to a profusion of consumer products. In traditional medicine, the whole fruit or its extracts are said to have numerous implied health effects, which remain scientifically unconfirmed as of 2014.

Cultivation

China

The majority of commercially produced wolfberries come from the Ningxia Hui Autonomous Region of north-central China and the Xinjiang Uyghur Autonomous Region of western China, where they are grown on plantations totaling 200,000 acres. In Zhongning County, Ningxia, wolfberry plantations typically range between 40 and 400 hectares (100–1000 acres or 500–6000 *mu*) in area. As of 2005, over 10 million mu have been planted with wolfberries in Ningxia.

Ningxia wolfberries have been cultivated along the fertile floodplains of the Yellow River for more than 700 years and have earned a reputation throughout Asia for premium quality sometimes described commercially as “red diamonds”. Government releases of annual wolfberry production, premium fruit grades, and export are based on yields from Ningxia, the region recognized with—

- the largest annual harvest in China, accounting for 45% of the nation’s total yield of wolfberries (50,000 tons, 2013);
- formation of an industrial association of growers, processors, marketers, and scholars of wolfberry cultivation to promote the berry’s commercial and export potential;
- the nation’s only source of superior grade wolfberries used by practitioners of traditional Chinese medicine.

Commercial volumes of wolfberries also grow in the Chinese regions of Inner Mongolia, Qinghai, Gansu, Shaanxi, Shanxi, and Hebei. When ripe, the oblong, red berries are tender and must be picked carefully or shaken from the vine into trays to avoid spoiling. The fruits are preserved by drying them in full sun on open trays or by mechanical dehydration employing a progressively increasing series of heat exposure over 48 hours.

Wolfberries are celebrated each August in Ningxia with an annual festival coinciding with the berry harvest. Originally held in Ningxia’s capital, Yinchuan, the festival has been based since 2000 in Zhongning County, an important center of wolfberry cultivation for the region. As Ningxia’s borders merge with three deserts, wolfberries are also planted to control erosion and reclaim irrigable soils from desertification.

China, the main supplier of wolfberry products in the world, had total exports generating US$120 million in 2004. This production derived from 82,000 hectare farmed nationwide, yielding 95,000 tons of wolfberries, which has increased from larger acreages cultivated in recent years.

Pesticide and Fungicide Use

China’s Green Food Standard, administered by the Chinese Ministry of Agriculture’s China Green Food Development Center, permits some pesticide and herbicide use.

United Kingdom

On June 18, 2007, the UK Food Standards Agency stated a significant history exists of the fruit being consumed in Europe before 1997, and has removed it from the Novel Foods list. It is now legal to sell the wolfberry in the UK as a food as reported by the Food Standards Agency, though with concerns over marketing claims over potential health benefits.

Importation of Mature Plants

Importation of wolfberry plants into the United Kingdom from most countries outside Europe is illegal, due to the possibility they could be vectors of diseases attacking Solanaceae crops, such as potato or tomato.

Canada and United States

During the first decade of the 21st century, farmers in Canada and the United States began cultivating goji on a commercial scale to meet potential markets for fresh berries, juice, and processed products.

Uses

Wolfberries are usually sold in open boxes and small packages in dried form.

Defrosted goji berries

Culinary

As a food, dried wolfberries are traditionally cooked before consumption. Dried wolfberries are often added to rice congee and almond jelly, as well as used in Chinese tonic soups, in combination with chicken or pork, vegetables, and other herbs such as wild yam, *Astragalus membranaceus*, *Codonopsis pilosula*, and licorice root. The berries are also boiled as a herbal tea, often along with chrysanthemum flowers and/or red jujubes, or with tea, and packaged teas are also available.

Various wines containing wolfberries (called *gǒuqí jiǔ* from) are also produced, including some that are a blend of grape wine and wolfberries.

Young wolfberry shoots and leaves are also harvested commercially as a leaf vegetable.

Medical Research

Although goji is the subject of basic research to determine if it has physiological properties, lack of clinical evidence and poor quality control in the manufacture of consumer products prevent goji from being clinically recommended or applied.

Safety Issues

In vitro testing has revealed that the tea inhibited warfarin metabolism, providing evidence for possible interaction between warfarin and undefined wolfberry phytochemicals.

Potentially harmful interactions may occur if wolfberry is consumed while taking other medications, such as those metabolised by the cytochrome P450 liver enzymes. Such drugs include warfarin, or drugs for diabetes or hypertension.

Marketing

Dried goji berries on sale in a market in France

Since the early 21st century, the dried fruit has been marketed in the West as a health food, often accompanied by scientifically unsupported claims regarding its purported health benefits.

Companies marketing the berries often propagate the unsupported claim that a Chinese man named Li Qing Yuen, who was said to have consumed wolfberries daily, lived to the age of 256 years (1677–1933).

Commercial Products Marketed Outside Asia

The presence of wolfberry in health food stores and grocery markets is increasing in the United Kingdom and other countries.

Other wolfberry consumer applications include:

- Dried berries (pictured above)
- Berry pieces in granola bars
- Yogurt products
- Green tea products

Commercial suppliers have processed wolfberry as

- An additive for manufacturing
- Juice concentrate
- Whole fruit purée

- Pulp powders
- Whole or ground seeds

No Special Conditions in the EU

In June 2007, the United Kingdom's Food Standards Agency (FSA) published the results of its inquiry concluding that goji berries were not a "novel" food in Europe. The inquiry found that the berries were already in wide use before 1997 and had a history sufficient to suggest safety. The FSA is an advisor to the European Union's European Food Safety Authority. Had the berries been found to be a novel food, goji berries would have required authorisation from the European Council and Parliament.

Marketing Claims Under Scrutiny in Canada and the United States

In January 2007, marketing statements for a goji juice product were the subject of an investigative report by CBC Television's consumer advocacy program *Marketplace*.

By one specific example in the CBC interview, Earl Mindell (then working for direct-marketing company *FreeLife International, Inc.*) falsely claimed the Memorial Sloan-Kettering Cancer Center in New York had completed clinical studies showing that use of wolfberry juice would prevent 75% of human breast cancer cases.

During 2006, the U.S. Food and Drug Administration (FDA) placed two goji juice distributors on notice with warning letters about unproven therapeutic benefits. These statements were in violation of the United States Food, Drug and Cosmetic Act [21 USC/321 (g)(1)] because they "establish the product as a drug intended for use in the cure, mitigation, treatment, or prevention of disease" when wolfberries or juice have had no such scientific evaluation. Additionally stated by the FDA, the goji juice was "not generally recognized as safe and effective for the referenced conditions" and therefore must be treated as a "new drug" under Section 21(p) of the Act. New drugs may not be legally marketed in the United States without prior approval of the FDA.

On May 29, 2009, a class action lawsuit was filed against FreeLife in the United States District Court of Arizona. This lawsuit alleges false claims, misrepresentations, false and deceptive advertising and other issues regarding FreeLife's Himalayan Goji Juice, GoChi, and TaiSlim products. This lawsuit seeks remedies for consumers who have purchased these products over the past several years.

Astragalus

Astragalus hamosus - MHNT

Astragalus is a large genus of about 3,000 species of herbs and small shrubs, belonging to the legume family Fabaceae and the subfamily Faboideae. It is the largest genus of plants in terms of described species. The genus is native to temperate regions of the Northern Hemisphere. Common names include milkvetch (most species), locoweed (in North America, some species) and goat's-thorn (*A. gummifer*, *A. tragacanthus*). Some pale-flowered vetches are similar in appearance, but vetches are more vine-like.

Ecology

Astragalus species are used as food plants by the larvae of some Lepidoptera species including many case-bearing moths of the genus *Coleophora*: *C. cartilaginella*, *C. colutella*, *C. euryaula*, and *C. onobrychiella* feed exclusively on *Astragalus*, *C. astragalella* and *C. gallipennella* feed exclusively on the species *Astragalus glycyphyllos*, and *C. hippodromica* is limited to *Astragalus gombo*.

Traditional Uses

The natural gum tragacanth is made from several species of *Astragalus* occurring in the Middle East, including *A. adscendens*, *A. gummifer*, *A. brachycalyx*, and *A. tragacanthus*. Also *Astragalus propinquus* (syn. *A. membranaceus*) has a history of use as a herbal medicine used in systems of traditional Chinese medicine and Persian medicine. In traditional Chinese medicine A. membranaceus has been used to reinforce qi and strengthen the superficial resistance, and promote the discharge of pus and the growth of new tissue.

Research

Biotechnology firms are working on deriving a telomerase activator from *Astragalus*. The chemical constituent cycloastragenol (also called TAT2) is being studied to help combat HIV, as well as infections associated with chronic diseases or aging. However, the National Institutes of Health states: "The evidence for using astragalus for any health condition is limited. High-quality clinical trials (studies in people) are generally lacking. There is some preliminary evidence to suggest that astragalus, either alone or in combination with other herbs, may have potential benefits for the immune system, heart, and liver, and as an adjunctive therapy for cancer".

Research at the UCLA AIDS Institute focused on the function of cycloastragenol in the aging process of immune cells, and its effects on the cells' response to viral infections. It appears to increase the production of telomerase, an enzyme that mediates the replacement of short bits of DNA known as telomeres, which play a key role in cell replication, including in cancer processes.

Supplement use

Extracts of *Astragalus propinquus* (syn. *A. membranaceus*) are marketed as life-prolonging extracts for human use. A proprietary extract of the dried root of *A. membranaceus*, called TA-65, "was associated with a significant age-reversal effect in the immune system, in that it led to declines in the percentage of senescent cytotoxic T cells and natural killer cells after six to twelve months of use". There are mixed data regarding *Astragalus*, its effects on telomerase, and cancer.

For example, although 80% of cancer cells utilize telomerase for their proliferation—a factor that might theoretically be exacerbated by *Astragalus*—the shortening of telomeres (resulting from such factors as stress and aging and possible contributors to malignancy), might also be mitigated by *Astragalus*. Thus, short telomeres result in chromosome instability, and the potential for telomere lengthening as a protection against cancer is possible. Additionally, scientists recently reported that cancer cells may proliferate precisely because of the lack of differentiation occurring via damaged or shortened telomere length. They propose that "forced" elongation of telomeres promotes the differentiation of cancer cells, probably reducing malignancy, which is strongly associated with a loss of cell differentiation.

Side Effects and Toxicology

Astragalus may interact with medications that suppress the immune system, such as cyclophosphamide. It may also affect blood sugar levels and blood pressure. Some *Astragalus* species can be toxic. For example, several species native to North America contain the alkaloid swainsonine, which may cause "locoism" in livestock. The toxicity of *Astragalus* taxa varies.

Patchouli

Patchouli (*Pogostemon cablin*) is a species of plant from the genus *Pogostemon*. It is a bushy herb of the mint family, with erect stems, reaching around 75 centimetres (2.5 ft) in height and bearing small, pale pink-white flowers. The plant is native to tropical regions of Asia, and is now extensively cultivated in China, Indonesia, Cambodia, Myanmar, India, Maldives, Malaysia, Mauritius, Seychelles, Madagascar, Taiwan, Philippines, Thailand, Vietnam, South America and the Caribbean.

Perfume

The heavy and strong scent of patchouli has been used for centuries in perfumes and, more recently, in incense, insect repellents, and alternative medicines. The word derives from the Tamil *patchai* (green), *ellai* (leaf). In Assamese it is known as *xukloti*. In Kannada it is known as *Pachhethene*.

Pogostemon cablin, *P. commosum*, *P. hortensis*, *P. heyneasus* and *P. plectranthoides* are all cultivated for their essential oil, known as patchouli oil.

Cultivation

Patchouli grows well in warm to tropical climates. It thrives in hot weather but not direct sunlight. If the plant withers due to lack of water, it will recover well and quickly after rain or watering. The seed-producing flowers are very fragrant and blossom in late fall. The tiny seeds may be harvested for planting, but they are very delicate and easily crushed. Cuttings from the mother plant can also be rooted in water to produce additional plants.

Extraction of Essential Oil

Extraction of patchouli's essential oil is by steam distillation of the dried leaves, requiring rupture of its cell walls by steam scalding, light fermentation, or drying. The main chemical component of patchouli oil is patchoulol, a sesquiterpine alcohol.

Leaves may be harvested several times a year and, when dried, may be exported for distillation. Some sources claim a highest quality oil is usually produced from fresh leaves distilled close to where they are harvested; others that boiling the dried leaves and fermenting them for a period of time is best.

Aroma Profile

- Germacrene-B
- Patchoulol
- Norpatchoulenol

Uses

Perfume

Patchouli is used widely in modern perfumery, by individuals who create their own scents, and in modern scented industrial products such as paper towels, laundry detergents, and air fresheners. Two important components of its essential oil are patchoulol and norpatchoulenol.

Insect Repellent

One study suggests that patchouli oil may serve as an all-purpose insect repellent. More specifically, the patchouli plant is claimed to be a potent repellent against the *Formosan subterranean termite.*

Incense

Patchouli is an important ingredient in East Asian incense. Both patchouli oil and incense underwent a surge in popularity in the 1960s and 1970s in the US and Europe, mainly as a result of the hippie movement of those decades.

Culinary

Patchouli leaves have been used to make an herbal tea. In some cultures, patchouli leaves are eaten as a vegetable or used as a seasoning.

Toys

In 1985 Mattel used patchouli oil in the plastic used to produce the action figure Stinkor in the Masters of the Universe line of toys.

Lingzhi Mushroom

The lingzhi mushroom or reishi mushroom (pinyin: *língzhī*; Japanese: rōmaji: *reishi*; Vietnamese: *linh chi*; literally: "soul/spirit mushroom") is a species complex that encompasses several fungal species of the genus *Ganoderma*, most commonly the closely related species *Ganoderma lucidum, Ganoderma tsugae,* and *Ganoderma lingzhi. G. lingzhi* enjoys special veneration in East Asia, where it has been used as a medicinal mushroom in traditional Chinese medicine for more than 2,000 years, making it one of the oldest mushrooms known to have been used medicinally.

Taxonomy and Naming

Names for the lingzhi fungus have a two thousand-year history. The Chinese term *lingzhi* was first recorded during the Han dynasty (206 BC – 9 AD). Petter Adolf Karsten named the genus *Ganoderma* in 1881.

Botanical Names

The fungus was given its first binomial name, *Boletus lucidus*, by English botanist William Curtis in 1781. The lingzhi's botanical names have Greek and Latin roots. The generic name *Ganoderma* derives from the Greek "brightness; sheen", hence "shining" and *derma* "skin". The specific epithet *lucidum* is Latin for "shining". *Tsugae* is derived from the Japanese word for "hemlock" (*tsuga*).

There are multiple species of lingzhi, scientifically known to be within the *Ganoderma lucidum* species complex and mycologists are researching the differences among species within this complex.

Chinese Names

In the Chinese language, *lingzhi* is made up of the compounds *ling* "spirit, spiritual; soul; miraculous; sacred; divine; mysterious; efficacious; effective" (cf. Lingyan Temple) and *zhi* "(traditional) plant of longevity; fungus; seed; branch; mushroom; excrescence". Fabrizio Pregadio notes, "The term *zhi*, which has no equivalent in Western languages, refers to a variety of supermundane substances often described as plants, fungi, or "excresences"." *Zhi* occurs in other Chinese plant names such as *zhima* "sesame" or "seed", and was anciently used a phonetic loan character for *zhi* "Angelica iris". Chinese differentiates *Ganoderma* species between *chizhi* "red mushroom" *G. lucidum* and *zizhi* "purple mushroom" *G. japonicum*.

Lingzhi has several synonyms. *ruicao* "auspicious plant" (with *rui* "auspicious; felicitous omen" and the suffix *cao* "plant; herb") is the oldest; the (c. 3rd century BCE) *Erya* dictionary defines *qiu* (interpreted as a miscopy of *jun* "mushroom") as *zhi* "mushroom" and the commentary of Guo Pu (276–324) says, "The [*zhi*] flowers three times in one year. It is a [*ruicao*] felicitous plant." Other Chinese names for *Ganoderma* include *ruizhi* "auspicious mushroom", *shenzhi* "divine mushroom" (with shen "spirit; god' supernatural; divine"), *mulingzhi* (with "tree; wood"), *xiancao* "immortality plant" (with xian "(Daoism) transcendent; immortal; wizard"), and *lingzhicao* or *zhicao* "mushroom plant".

Since both Chinese *ling* and *zhi* have multiple meanings, *lingzhi* has diverse English translations. Renditions include "[zhi] possessed of soul power", "Herb of Spiritual Potency" or "Mushroom of Immortality", "Numinous Mushroom", "divine mushroom", "divine fungus", "Magic Fungus", and "Marvelous Fungus".

Japanese Names

Japanese language *Reishi* is a Sino-Japanese loan word from *lingzhi*. This modern Japanese kanji is the shinjitai "new character form" for the kyūjitai "old character form" .

Reishi synonyms divide between Sino-Japanese borrowings and native Japanese coinages. Sinitic loanwords include literary terms such as *zuisō* (from *ruicao*) "auspicious plant" and *sensō* (from *xiaocao*) "immortality plant". A common native Japanese name is *mannentake* "10,000 year mushroom". The Japanese writing system uses *shi* or *shiba* for "grass; lawn; turf" and *take* or *kinoko* for "mushroom" (e.g., shiitake). Other Japanese terms for *reishi* include *kadodetake* "departure mushroom", *hijiridake* "sage mushroom", and *magoshakushi* "grandchild ladle".

Korean Names

Korean language *Yeong Ji* or *Yung Gee* is a word from hanja of *lingzhi*. It is also called *seon-cho* , *gil-sang-beo-seot* , *yeong ji cho* or *jeok ji*. It can be classified by its color such as *ja-ji* for purple one, *heuk-ji* for black, *cheong-ji* for blue or green, *baek-ji* for white, *hwang-ji* for yellow.

Vietnamese Names

The Vietnamese language *linh chi* is a Chinese loanword used in tiếng Việt. It is often used with the Vietnamese word for mushroom nấm (nấm Linh Chi) which is the equivalent of *Ganoderma lucidum* or reishi mushroom.

English Names

English *lingzhi* or *ling chih* (sometimes spelled "*ling chi*" from French EFEO Chinese transcription) is a Chinese loanword. The *Oxford English Dictionary* gives Chinese "*líng* divine + *zhī* fungus" as the origin of *ling chih* or *lingzhi*, and defines, "The fungus *Ganoderma lucidum*, believed in China to confer longevity and used as a symbol of this on Chinese ceramic ware." The *OED* notes the earliest recorded usage of the Wade–Giles romanization *ling chih* in 1904, and of the Pinyin *lingzhi* in 1980. In addition to the transliterated loanword, English names include "glossy ganoderma" and "shiny polyporus".

Description

Young sporocarp

Lingzhi is a polypore mushroom that is soft (when fresh), corky, and flat, with a conspicuous red-varnished, kidney-shaped cap and, depending on specimen age, white to dull brown pores underneath. It lacks gills on its underside and releases its spores through fine pores, leading to its morphological classification as a polypore.

Varieties

Grown Reishi Mushroom

With the advent of genome sequencing, the genus *Ganoderma* has been undergoing taxonomic reclassification. Prior to genetic analyses of fungi, they were classified according to their morphological characteristics. The ITS region of the Ganoderma genome is considered to be a standard barcode marker. It was once thought that *Ganoderma lucidum* generally occurs in two growth forms- a large specimen with a small or nonexistent stalk found in North America and is sessile, and the other a smaller specimen with a long, narrow stalk found mainly in the tropics. However recent molecular evidence has shown the first stalkless form to be a distinct species called *G.sessile*, a name given to North American specimens by William Alfonso Murrill in 1902. Environmental conditions play a substantial role in the lingzhi's manifest morphological characteristics. For example, elevated carbon dioxide levels result in stem elongation in lingzhi. Other formations include antlers without a cap; these may be affected by carbon dioxide levels as well. The species were also formerly differentiated by their colors, of which the red reishi was considered to be the most researched. The three main factors that most greatly influence fruitbody development morphology are light, temperature, and humidity. (Water and air quality are impactful but to a lesser degree).

Depending on the environmental or cultivation conditions
the reishi can lose its umbrella shape and resemble antlers instead.

Biochemistry

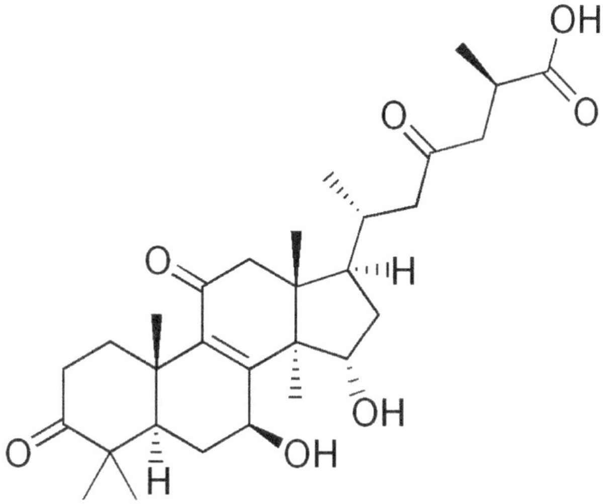

Ganoderic acid A, a compound isolated from lingzhi

Ganoderma lucidum produces a group of triterpenes, called ganoderic acids, which have a molecular structure similar to steroid hormones. It also contains other compounds often found in fungal materials, including polysaccharides (such as beta-glucan), coumarin, mannitol, and alkaloids. Sterols isolated from the mushroom include, ganoderol, ganoderenic acid, ganoderiol, ganodermanontriol, lucidadiol, and ganodermadiol. Fungal immunomodulatory proteins (FIPs) are bioactive ingredients within genera *Ganoderma* that have immune building properties. FIPs stimulate different cells and cellular components that enable immune response. Some of the molecules and cells that FIPs influence include T and B lymphocytes, natural killer cells, macrophages, and regulation of human monocyte DCs. Improved function of these cells will promote cytokine expression such as IL2, IL-4, and IFN-gamma which will support lymphocyte proliferation, immune cell initiation, and tumor inhibiting factors. Immunostimulation from FIPs could cause increased expression of costimulatory molecules and major histocompatibility complexes (MHC) which help to create a pathway for immune response to take place.

Habitat

Ganoderma lucidum, and its close relative *Ganoderma tsugae*, grow in the northern Eastern Hemlock forests. These two species of bracket fungus have a worldwide distribution in both tropical and temperate geographical regions, growing as a parasite or saprotroph on a wide variety of trees. Similar species of *Ganoderma* have been found growing in the Amazon. In nature, lingzhi grows at the base and stumps of deciduous trees, especially maple. Only two or three out of 10,000 such aged trees will have lingzhi growth, and therefore its wild form is extremely rare. Today, lingzhi is effectively cultivated on hardwood logs or sawdust/woodchips.

History

The Chinese classics first used *zhi* during the Warring States period (475–221 BCE) and *lingzhi* during the Han Dynasty (206 BCE-220 CE).

The word *zhi* occurs approximately 100 times in classical texts. Occurrences in early Chinese histories, such as the (91 BCE) *Shiji* "Records of the Grand Historian" and (82 CE) *Hanshu* "Book of Han", predominantly refer to the "Mushroom of Immortality; elixir of life". They record that fangshi "masters of esoterica; alchemists; magicians", supposedly followers of Zou Yan (305–240 BCE), claimed to know secret locations like Mount Penglai where the magic *zhi* mushroom grew. Some sinologists propose that the mythical *zhi* derived from Indian legends about soma that reached China around the 3rd century BCE. *Fangshi* courtiers convinced Qin and Han emperors, most notably Qin Shi Huang (r. 221–210 BCE) and Emperor Wu of Han (r. 141–87 BCE), to dispatch large expeditions (e.g., Xu Fu in 219 BCE) seeking the *zhi* Plant of Immortality, but none produced tangible results. *Zhi* occurrences in other classical texts often refer to an edible fungus. The *Liji* "Record of Ritual" lists *zhi* "lichens" as a type of condiment. The *Chuci* "Song of the South" metaphorically mentions, "The holy herb is weeded out". The *Huainanzi* "Philosophers of Huainan" records a *zizhi* "Purple Mushroom" Aphorism, "The *zhi* fungus grows on mountains, but it cannot grow on barren boulders."

Man holding *ganoderma* by Chen Hongshou

The word *lingzhi* was first recorded in a *fu* "rhapsody; prose-poem" by the Han dynasty polymath Zhang Heng (CE 78–139). His *Xijing fu* "Western Metropolis Rhapsody" description of Emperor Wu of Han's (104 BCE) Jianzhang Palace parallels *lingzhi* with *shijun* "rock mushroom": "Raising huge breakers, lifting waves, That drenched the stone mushrooms on the high bank, And soaked the magic fungus on vermeil boughs." The commentary by Xue Zong (d. 237) notes these fungi were eaten as drugs of immortality.

The (ca. 1st–2nd century CE) *Shennong bencao jing* "Divine Farmer's Classic of Pharmaceutics" classifies *zhi* into six color categories, each of which is believed to benefit the *qi* "Life Force" in a different part of the body: *qingzhi* "Green Mushroom" for Liver, *chizhi* "Red Mushroom" for heart, *huangzhi* "Yellow Mushroom" for spleen, *baizhi* "White Mushroom" for Lung, *heizhi* "Black Mushroom" for kidney, and *zizhi* "Purple Mushroom" for Essence. Commentators identify this red *chizhi* (or *danzhi* "cinnabar mushroom") as the *lingzhi*.

Chi Zhi (*Ganoderma rubra*) is bitter and balanced. It mainly treats binding in the chest, boosts the heart qi, supplements the center, sharpens the wits, and [causes people] not to forget [i.e., improves the memory]. Protracted taking may make the body light, prevent senility, and prolong life so as to make one an immortal. Its other name is *Dan Zhi* (Cinnabar Ganoderma). It grows in mountains and valleys.

While Chinese texts have recorded medicinal uses of *lingzhi* for more than 2,000 years, a few sources erroneously claim more than 4,000 years. Modern scholarship accepts neither the historicity of Shennong "Divine Farmer" (legendary inventor of agriculture, traditionally r. 2737–2697 BCE) nor that he wrote the *Shennong bencao jing*.

The (ca. 320 CE) *Baopuzi*, written by the Jin Dynasty Daoist scholar Ge Hong, has the first classical discussion of *Zhi*. Based upon no-longer extant texts, Ge distinguishes five categories of *zhi*, each with 120 varieties: *Shizhi* "stone *Zhi*", *Muzhi* "wood *Zhi*", *Caozhi* "Plant *Zhi*", *Rouzhi* "flesh *zhi*", and *junzhi* "mushroom *zhi*. For example, the "mushroom *zhi*".

Tiny excresences. These grow deep in the mountains, at the base of large trees or beside springs. They may resemble buildings, palanquins and horses, dragon and tigers, human beings, or flying birds. They may be any of the five colors. They too number 120 for which there exist illustrations. All are to be sought and gathered while using Yu's Pace, and they are to be cut with a bone knife. When dried in the shade, powdered, and taken by the inch-square spoonful, they produce geniehood. Those of the intermediate class confer several thousands of years, and those of the lowest type a thousand years of life.

Yu's Pace is a Daoist ritual walking technique. Pregadio concludes, "While there may be no better term than "mushrooms" or "excresences" to refer to them, and even though Ge Hong states that they "are not different from natural mushrooms (*ziran zhi*) (*Baopuzi* 16.287)", the *Zhi* pertain to an intermediate dimension between mundane and transcendent reality."

The (1596) *Bencao Gangmu* ("Compendium of Materia Medica") has a *zhi* category that includes six types of *Zhi* (calling the green, red, yellow, white, black, and purple ones from the *Shennong bencao jing* the *liuzhi* "six mushrooms") and sixteen other fungi, mushrooms, and lichens (e.g., *mu'er* "wood ear" "Cloud ear fungus; *Auricularia auricula-judae*"). The author Li Shizhen classified these six differently colored *Zhi* as *Xiancao* "immortality herbs", and described the effects of *Chizhi* "red mushroom":

It positively affects the life-energy, or *Qi* of the heart, repairing the chest area and benefiting those with a knotted and tight chest. Taken over a long period of time, agility of the body will not cease, and the years are lengthened to those of the Immortal Fairies.

Stuart and Smith's classical study of Chinese herbology describes the *zhi*.

(Chih) is defined in the classics as the plant of immortality, and it is therefore always considered to be a felicitous one. It is said to absorb the earthy vapors and to leave a heavenly atmosphere. For this reason it is called (Ling-chih.) It is large and of a branched form, and probably represents Clavaria or Sparassis. Its form is likened to that of coral.

The *Bencao Gangmu* does not list *lingzhi* as a variety of *zhi*, but as an alternate name for the *shi'er* "stone ear" "Umbilicaria esculenta" lichen. According to Stuart and Smith,

[The Shih-erh is] edible, and has all of the good qualities of the (Chih), being also used in the treatment of gravel, and said to benefit virility. It is specially used in hemorrhage from the bowels and prolapse of the rectum. While the name of this would indicate that it was one of the Auriculariales, the fact that the name (Ling-chih) is also given to it might place it among the Clavariaceae.

Chinese pharmaceutical handbooks on *Zhi* mushrooms were the first illustrated publications in the history of mycology. The historian of Chinese science Joseph Needham discussed a no-longer extant Liang Dynasty (502–587) illustrated text called *Zhong Shenzhi* "On the Planting and Cultivation of Magic Mushrooms".

The pictures of mushrooms in particular must have been an extremely early landmark in the history of mycology, which was a late-developing science in the West. The title of [this book] shows that fungi of some kind were being regularly cultivated – hardly as food, with that special designation, more probably medicinal, conceivably hallucinogenic."

The (1444) Ming Dynasty edition *Daozang* "Daoist canon" contains the *Taishang lingbao zhicao pin* "Classifications of the Most High Divine Treasure Mushroom Plant", which categorizes 127 varieties of *Zhi*. A (1598) Ming reprint includes woodblock pictures.

In Chinese art, the *lingzhi* symbolizes great health and longevity, as depicted in the imperial Forbidden City and Summer Palace. It was a talisman for luck in the traditional culture of China, and the goddess of healing Guanyin is sometimes depicted holding a *reishi* mushroom.

Preparation

Because of its bitter taste, lingzhi is traditionally prepared as a hot water extract product. Thinly sliced or pulverized lingzhi (either fresh or dried) is added to a pot of boiling water, the water is then reduced to a simmer, and the pot is covered; the lingzhi is then simmered for two hours. The resulting liquid is fairly bitter in taste and dark, with the more active red lingzhi more bitter than the black. The process is sometimes repeated for additional concentration. Alternatively, it can be used as an ingredient in a formula decoction or used to make an extract (in liquid, capsule, or powder form). The more active red forms of lingzhi are far too bitter to be consumed in a soup.

Medical uses

Ganoderma lucidum is widely used and recommended by Asian physicians and naturopaths for its supporting effects on the immune system. Preliminary laboratory research and limited preclinical trials have suggested that *G. lucidum* has potential anticancer and immunomodulatory properties, and cancer patients have been using *G. lucidum*-derived supplements more commonly as an alternative medicine. However, a 2015 Cochrane database review found insufficient evidence to justify the use of *G. lucidum* as a first-line cancer treatment. It suggests that *G. lucidum* may have "benefit as an alternative adjunct to conventional treatment in consideration of its potential of enhancing tumour response and stimulating host immunity".

Traditional African Medicine

Traditional African medicine is an alternative medicine discipline involving indigenous herbalism and African spirituality, typically involving diviners, midwives, and herbalists. Practitioners of traditional African medicine claim to be able to cure various and diverse conditions such as cancers, psychiatric disorders, high blood pressure, cholera, most venereal diseases, epilepsy, asthma, eczema, fever, anxiety, depression, benign prostatic hyperplasia, urinary tract infections, gout, and healing of wounds and burns and even Ebola.

Diagnosis is reached through spiritual means and a treatment is prescribed, usually consisting of a herbal remedy that is considered to have not only healing abilities but also symbolic and spiritual significance. Traditional African medicine, with its belief that illness is not derived from chance occurrences, but through spiritual or social imbalance, differs greatly from modern scientific medicine, which is technically and analytically based. In the 21st century, modern pharmaceuticals and medical procedures remain inaccessible to large numbers of African people due to their relatively high cost and concentration of health facilities in urban centres.

Before the establishment of science-based medicine, traditional medicine was the dominant medical system for millions of people in Africa but the arrival of the Europeans was a noticeable turning point in the history of this ancient tradition and culture. Herbal medicines in Africa are generally not adequately researched, and are weakly regulated. There is a lack of the detailed documentation of the traditional knowledge, which is generally transferred orally. Serious adverse effects can result from misidentification or misuse of healing plants.

The geographical reach of this article is sub-saharan Africa. Though, of course neighbouring medical traditions influenced traditional *African medicine.*

History

Colonial Era

Science has, in the past, considered methods of traditional knowledge as primitive and backward. Under colonial rule, traditional diviner-healers were outlawed because they were considered by many nations to be practitioners of witchcraft and magic, and declared illegal by the colonial authorities, creating a war against aspects of the indigenous culture that were seen as witchcraft.

During this time, attempts were also made to control the sale of herbal medicines. After Mozambique obtained independence in 1975, attempts to control traditional medicine went as far as sending diviner-healers to re-education camps. As colonialism and Christianity spread through Africa, colonialists built general hospitals and Christian missionaries built private ones, with the hopes of making headway against widespread diseases. Little was done to investigate the legitimacy of these practices, as many foreigners believed that the native medical practices were pagan and superstitious and could only be suitably fixed by inheriting Western methods according to Onwuanibe. During times of conflict, opposition has been particularly vehement as people are more likely to call on the supernatural realm. Consequently, doctors and health practitioners have, in most cases, continued to shun traditional practitioners despite their contribution to meeting the basic health needs of the population.

Modern Period

Nurse at Koidu Hospital in Sierra Leone consulting with patients.

In recent years, the treatments and remedies used in traditional African medicine have gained more appreciation from researchers in science. Developing countries have begun to realize the high costs of modern health care systems and the technologies that are required, thus proving Africa's dependence to it. Due to this, interest has recently been expressed in integrating traditional African medicine into the continent's national health care systems. An African healer embraced this concept by making a 48-bed hospital, the first of its kind, in Kwa-Mhlanga, South Africa, which combines traditional methods with homeopathy, iridology, and other Western healing methods, even including some traditional Asian medicine. However, the highly sophisticated technology involved in modern medicine, which is beginning to integrate into Africa's health care system, could possibly destroy Africa's deep-seated cultural values.

Diagnostics

The diagnoses and chosen methods of treatment in traditional African medicine rely heavily on spiritual aspects, often based on the belief that psycho-spiritual aspects should be addressed before medical aspects. In African culture, it is believed that "nobody becomes sick without sufficient reason." Traditional practitioners look at the ultimate "who" rather than the "what" when locating the cause and cure of an illness, and the answers given come from the cosmological beliefs of the people. Rather than looking to the medical or physical reasons behind an illness, traditional healers attempt to determine the root cause underlying it, which is believed to stem from a lack of balance between the patient and his or her social environment or the spiritual world, not by natural causes. Natural causes are, in fact, not seen as natural at all, but manipulations of spirits or the gods. For example, sickness is sometimes said to be attributed to guilt by the person, family, or village for a sin or moral infringement. The illness, therefore, would stem from the displeasure of the gods or God, due to an infraction of universal moral law. According to the type of imbalance the individual is experiencing, an appropriate healing plant will be used, which is valued for its symbolic and spiritual significance as well as for its medicinal effect.

When a person falls ill, a traditional practitioner uses incantations to make a diagnosis. Incantations are thought to give the air of mystical and cosmic connections. Divination is typically used if the illness is not easily identified, otherwise, the sickness may be quickly diagnosed and given a remedy. If divination is required, then the practitioner will advise the patient to consult a diviner who can further give a diagnosis and cure. Contact with the spirit world through divination often requires not only medication, but sacrifices.

Treatments

Traditional practitioners use a wide variety of treatments ranging from "magic" to biomedical methods such as fasting and dieting, herbal therapies, bathing, massage, and surgical procedures. Migraines, coughs, abscesses, and pleurisy are often treated using the method of "bleed-cupping" after which an herbal ointment is applied with follow-up herbal drugs. Animals are also sometimes used to transfer the illness to afterward. Some cultures also rub hot herbal ointment across the patient's eyelids to cure headaches. Malaria is treated by both drinking and using the steam from an herbal mixture. Fevers are often treated using a steam bath. Also, vomiting is induced, or emetics, in an attempt to cure some diseases. For example, raw beef is soaked in the drink of an alcoholic person to induce vomiting and nausea and treat alcoholism. In the Bight of Benin, the natives have been known to use the fat of a boa constrictor to allegedly cure gout and rheumatism, and it also is thought to relieve chest pain when rubbed into the skin. Approximately 60%-80% of the people in Africa rely on traditional remedies to treat themselves for various diseases. A large percentage of the people in South Africa also rely on traditional remedies to treat their animals for various diseases.

Medicinal Plants

Africa is endowed with many plants that can be used for medicinal purposes to which they have taken full advantage. In fact, out of the approximated 6400 plant species used in tropical Africa, more than 4000 are used as medicinal plants. Medicinal plants are used in the treatments of many diseases and illnesses, the uses and effects of which are of growing interest to Western societies. Not only are plants used and chosen for their healing abilities, but they also often have symbolic and spiritual significance. For example, leaves, seeds, and twigs that are white, black and red are seen as especially symbolic or magical and possess special properties. Examples of some medicinal plants include:

Prunus africana with stripped bark.

- Pygeum (*Prunus africana*): Pygeum is not only used in traditional African medicine, but has developed a following around the world, as a cure for mild-to-moderate benign prostatic hyperplasia, claimed by its users to increase the ease of urination and reduce inflammation and cholesterol deposits. In traditional African practice, the bark is made into tea, whereas elsewhere in the world it is found in powders, tinctures, and pills. Pygeum has been sold in Europe since the 1970s and is harvested in mass quantities in Cameroon and Madagascar each year.
- *Securidaca longipedunculata:* This is a tropical plant found almost everywhere across the continent with different uses in every part of Africa. In Tanzania, the dried bark and root are used as a laxative for nervous system disorders, with one cup of the mixture being taken daily for two weeks. In East Africa, dried leaves from the plant are used in the treatment of wounds and sores, coughs, venereal diseases, and snakebites. In Malawi, the leaves are also used for wounds, coughs, venereal diseases, and snakebites, as well as bilharzia, and the dried leaves are used to cure headaches. In other parts of the continent, parts of the plant are used to cure skin diseases, malaria, impotence, epilepsy, and are also used as an aphrodisiac.

Preparing and drying out freshly dug traditional medicines (muti)

A study, entitled *ACE Inhibitor Activity of Nutritive Plants in Kwa-Zulu Natal*, was conducted by Irene Mackraj and S. Ramesar, both of the Department of Physiology and Physiological Chemistry; and H. Baijnath, Department of Biological and Conservation Sciences; University of Kwa-Zulu Natal, Durban, South Africa to examine the effectiveness of 16 plants growing in Africa's KwaZulu-Natal region, concluding that eight plant extracts may hold value for treating high blood pressure (hypertension). The plants (known locally as muti) used by traditional healers that the team examined were:

Plant	Description
Amaranthus dubius	a flowering plant, also known as spleen amaranth
Amaranthus hybridus	commonly known as smooth pig-weed or slim amaranth
Amaranthus spinosus	also known as spiny amaranth
Asystasia gangetica	an ornamental, ground cover known as Chinese violet. Also used in Nigerian folk medicine for the management of asthma.
Centella asiatica	a small herbaceous annual plant commonly referred to as Asiatic pennywort
Ceratotheca triloba	a tall annual plant that flowers in summer sometimes referred to as poppy sue
Chenopodium album	also called lamb's quarters, this is a weedy annual plant
Emex australis	commonly known as southern three corner jack
Galinsoga parviflora	commonly referred to as gallant soldier
Justicia flava	also known as yellow *justicia* and taken for coughs and treatment of fevers
Momordica balsamina	an African herbal traditional medicine also known as the balsam apple
Oxygonum sinuatum	an invasive weed with no common name
Physalis viscosa	known as starhair ground cherry
Senna occidentalis	a very leafy tropical shrub whose seeds have been used in coffee; called septic weed
Solanum nodiflorum	also known as white nightshade
Tulbaghia violacea	a bulbous plant with hairless leaves often referred to as society or wild garlic

Of the 16 plants, *Amaranthus dubius*, *Amaranthus hybridus*, *Asystasia gangetica*, *Galinsoga parviflora*, *Justicia flava*, *Oxygonum sinuatum*, *Physalis viscosa*, and *Tulbaghia violacea* were found to have some positive effects, with the latter proving to be the most promising with the ability to lower one's blood pressure. The unique infrageneric diversity of *Aloe L.* (*Aloaceae/Asphodelaceae*) and the broad therapeutic uses in southern Africa, suggest its cultural significance in the subcontinent. Palms are very commonly used as ritual elements, as well as an ingredient in ATM.

Recent Discoveries

- Other plants scientifically studied in 2016 include *Erigeron floribundus*, a medicinal plant used in Cameroon, inhibited *Staphylococcus aureus* nicotinate mononucleotide adenylyltransferase (NadD) yet had no effect the human analogue. Among the constituents of the essential oil are spathulenol and limonene.

- A traditional medicine from the tropical Olon tree, and another species of *Zanthoxylum* genus, was found to have synergistic compounds that kill both mosquitoes and their plasmodium parasites.

Spirituality

Famous Bedik diviner outside Iwol, southeast Senegal (West Africa).
He predicted outcomes by examining the color of the organs of sacrificed chickens.

Some healers may employ the use of charms, incantations, and the casting of spells in their treatments. The dualistic nature of traditional African medicine between the body and soul, matter, and spirit and their interactions with one another are also seen as a form of magic. Richard Onwuanibe gives one form of magic the name "Extra-Sensory-Trojection." This is the belief among the Ibos of Nigeria that medicine men can implant something into a person from a distance to inflict sickness on them. This is referred to by the Ibos as*egba ogwu*. To remove the malignant object, the intervention of a second medicine man is typically required, who then removes it by making an incision in the patient. *Egba ogwu* involves psychokinetic processes. Another form of magic used by these practitioners, which is more widely known, is sympathetic magic, in which a model is made of the victim. Actions performed on the model are transferred to the victim, in a manner similar to the familiar voodoo doll. "In cases where spirits of deceased relatives trouble the living and cause illness, medicine men prescribe remedies, often in the form of propitiatory sacrifice, in order to put them to rest so that they will no longer trouble the living, especially children." Using charms and amulets to cure diseases and illnesses is an uncertain and clouded practice that requires more scientific investigation.

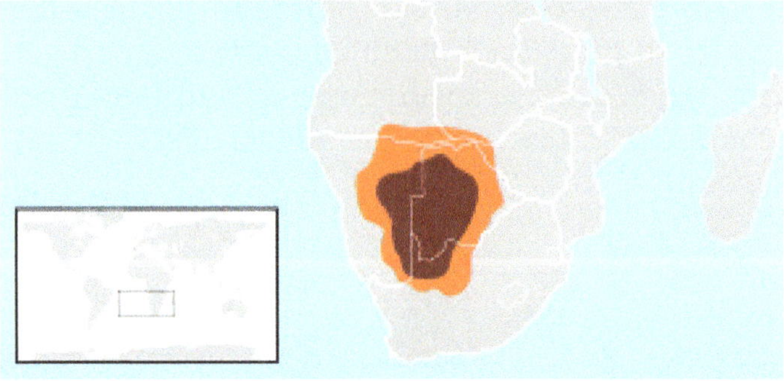

The Kalahari Desert (shown in red) and the surrounding Kalahari Basin (in orange).

In African cultures, the act of healing is considered a religious act. Therefore, the healing process often attempts to appeal to God because it is ultimately God who can not only inflict sickness, but provide a cure. Africans have a religious world view which makes them aware of the feasibility of divine or spirit intervention in healing with many healers referring to the supreme god as the source of their medical power. For example, the !Kung people of the Kalahari Desert believe that the great God Hishe created all things and, therefore, controls all sickness and death. Hishe, however, bestows mystical powers for curing sickness on certain men. Hishe presents himself to these medicine men in dreams and hallucinations, giving them curative power. Because this god is generous enough to give this power to the medicine men, they are expected to practice healing freely. The !Kung medicine men effect a cure by performing a tribal dance. Loma Marshall, who took expeditions to South West Africa with her family to study the !Kung people, writing two books on their findings, describes the ceremonial curing dance as follows:

At the dances not only may the sick be cured, but pending evil and misfortune averted. The !Kung believe that the great god may send *Gauwa* or the *gauwas* at any time with ill for someone and that these beings may be lurking awaiting their chance to inflict it. The medicine men in the dances combat them, drive them away, and protect the people. Usually there are several medicine men performing at the same time. To cure they go into trance, which varies in depth as the ceremony proceeds... When a man begins, he leaves the line of dancing men, and still singing, leans over the person he is going to cure, going eventually to every person present, even the infants. He places one hand on the person's chest, one on his or her back, and flutters his hands. The !Kung believe that in this way he draws the sickness, real or potential, out of the person through his own arms into himself... Finally, the medicine man throws up his arms to cast the sickness out, hurling it into the darkness back to *Gauwa* or the *gauwasi*, who are there beyond the firelight, with a harp, yelping cry of "Kai Kai Kai."

Loma Marshall does not give any information as to whether or not the dance is successful in curing the patient but says that it purges the people's emotions for their "support and solace and hope."

Traditional Medicinal Practitioners

Inyanga/Sangoma from Johannesburg, South Africa

Many traditional medicinal practitioners are people without education, who have rather received knowledge of medicinal plants and their effects on the human body from their forebears. They have a deep and personal involvement in the healing process and protect the therapeutic knowledge by keeping it a secret.

In a manner similar to orthodox medicinal practice, the practitioners of traditional medicine specialize in particular areas of their profession. Some, such as the inyangas of Swaziland are experts in herbalism, whilst others, such as the South African sangomas, are experts in spiritual healing as diviners, and others specialize in a combination of both forms of practice. There are also traditional bone setters and birth attendants. Herbalists are becoming more and more popular in Africa with an emerging herb trading market in Durban that is said to attract between 700,000 and 900,000 traders per year from South Africa, Zimbabwe, and Mozambique. Smaller trade markets exist in virtually every community. Their knowledge of herbs has been invaluable in African communities and they were the only ones who could gather them in most societies. Midwives also make extensive use of indigenous plants to aid childbirth. African healers commonly "describe and explain illness in terms of social interaction and act on the belief that religion permeates every aspect of human existence."

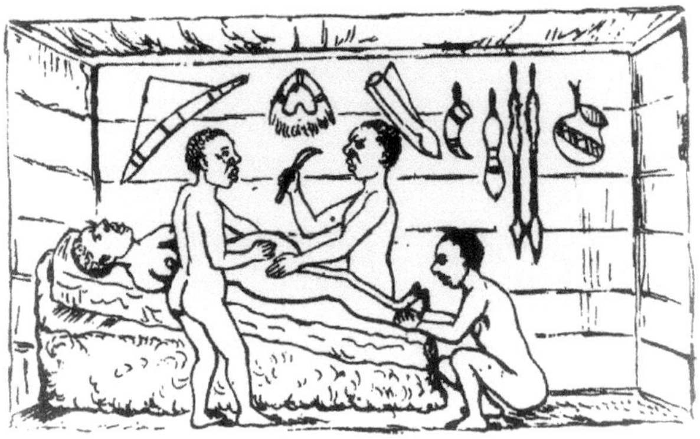

Successful Cesarean section performed by indigenous healers in Kahura, Uganda. As observed by R. W. Felkin in 1879.

Payments

Traditional healers, like any other profession, are rewarded for their services. In African societies, the payment for a treatment depends on its efficacy. They do not request payment until after the treatment is given. This is another reason many prefer traditional healers to western doctors who require payment before the patient has assessed the effectiveness of the treatment. The payment methods have changed over time, with many practitioners now asking for monetary payment, especially in urban settings, rather than their receiving good in exchange, as happened formerly.

There are also a growing number of fraudulent practitioners who only interested in making money, especially in urban areas.

Learning the Trade

Some healers learn the trade through personal experience while being treated as a patient who decide to become healers upon recovery. Others become traditional practitioners through a "spiritual calling" and, therefore, their diagnoses and treatments are decided through the supernatural. In some cultures, a sign of calling can come from mental disarrangement said to be caused by *agwu Nshi,* the spirit of divining, through which the healer gains inspiration. Through this training, psychological stability is eventually attained. Another route is receive the knowledge and skills passed down informally from a close family member such as a father or uncle, or even a mother or aunt in the case of midwives. Apprenticeship to an established practitioner, who formally teaches the trade over a long period of time and is paid for their tutoring, is another route to becoming a healer. The training is complex, depending on the kind of medical practice that the aspiring practitioner wants to be a part of. Once the trainee is officially initiated as a healer, they are, in some societies, considered to be half-man and half spirit, possessing the power to mediate between the human and supernatural world to invoke spiritual power in their healing processes.

Importance

In Africa, the importance of traditional healers and remedies made from indigenous plants play a crucial role in the health of millions. According to the International Development Research Centre (IDRC), one estimate puts the number of Africans who routinely use these services for primary health care as high as 85% in Sub-Saharan Africa. The relative ratios of traditional practitioners and university trained doctors in relation to the whole population in African countries showcases this importance. For example, in Ghana, in Kwahu district, for every traditional practitioner there are 224 people, against one university trained doctor for nearly 21,000. In Swaziland, the same situation applies, where for every healer there are 110 people whereas for every university trained doctor there are 10,000 people. According to Nairobi-based specialist in biodiversity and traditional medicine with the IDRC Francois Gasengayire, there is one healer for every 200 people in the Southern Africa region which is a much greater doctor-to-patient ratio than is found in North America.

Ratios of doctors (practicing modern medicine) and traditional medical practitioners to patients in east and southern Africa:

Country	Doctor:Patient	TMP:Patient	References
Botswana		TMPs estimated at 2,000 in 1990	Moitsidi, 1993
Eritrea	Medical doctors estimated at 120 in 1995		Government of Eritrea, 1995
Ethiopia	1:33,000		World Bank, 1993
Kenya	1:7,142 (overall)	1:987 (Urban-Mathare)	World Bank, 1993
	1:833 (Urban-Mathare)	1:378 (Rural-Kilungu)	Good. 1987
Lesotho		Licensed TMPs estimated at 8,579 in 1991	Scott *et al.* 1996
Madagascar	1:8,333		World Bank, 1993
Malawi	1:50,000	1:138	Msonthi and Seyani, 1986
Mozambique	1:50,000	1:200	Green *et al.* 1994
Namibia		1:1,000 (Katutura) 1:500 (Cuvelai) 1:300(Caprivi)	Lumpkin, 1994
Somalia	1:14,285 (Overall) 1:2,149 (Mogadishu) 1:54,213 (Central region) 1:216,539 (Sanag)		World Bank, 1993; Elmi *et al.* 1983

South Africa	1:1,639 (Overall) 1:17,400 (Homeland areas)	1:700-1,200 (Venda)	World Bank, 1993 (Venda and Overall), Savage, 1985* Arnold and Gulumian, 1987* (Homeland areas)
Sudan	1:11,000		World Bank, 1993
Swaziland	1:10,000	!:100	Green, 1985; Hoff and Maseko, 1986
Tanzania	1:33,000	1:350-450 in DSM	World Bank, 1993; Swantz, 1984
Uganda	1:25,000	1:708	World Bank, 1993; Amai, 1997
Zambia	1:11,000		World Bank, 1993
Zimbabwe	1:6,250	1:234 (urban) 1:956(rural)	World Bank, 1993; Gelfand *et al.* 1985

*TMP refers to Traditional Medical Practitioner

This table showing the ratio of traditional medical practitioner to patient and Western practitioner to patient shows that in many parts of Africa, practitioners trained in modern medicine are few and far between. Because of this, healers prove to be a large and influential group in primary health care and an integral part of the African culture and are required for the health of its people. Without them, many people would go untreated.

Medications and treatments that Western pharmaceutical companies manufacture are far too costly and not available widely enough for most Africans. Many rural African communities are not able to afford the high price of pharmaceuticals and can not readily obtain them even if they were affordable; therefore, healers are their only means of medical help. According to Sekagya Yahaya Hills, who is a university-trained dentist and a traditional healer in Uganda, there are promising signs that some of the plant-based remedies offered by medicine-men are not just affordable, but also effective, even in treating AIDS. Hills read his Declaration of Traditional Healers at the 13th International Conference on AIDS and sexually transmitted infections (STIs) in Africa, which summarized the important role of traditional medicine, stating: "As traditional healers, we are the most trusted and accessible health care providers in our communities. We have varied and valuable experience in treating AIDS-related illness and accept the great responsibility of continuing to do so." Because this form of medicine is "the most affordable and accessible system of health care for the majority of the African rural population," the African Union declared 2001 to 2010 to be the Decade for African Traditional Medicine with the goal of making "safe, efficacious, quality, and affordable traditional medicines available to the vast majority of the people."

Excessive use of plants is an oecologic risk, as this may cause their extinction.

Relationship with Modern Medicine

Before the establishment of science-based medicine, traditional medicine was the dominant medical system for millions of people in Africa but the arrival of the Europeans was a noticeable turning point in the history of this ancient tradition and culture. Although modern scientific medicine is successful in developed countries, it doesn't have the same positive impact in many of the underdeveloped African countries. Though Western practices can make an impact in health care practices, in certain areas such as in the spread of various diseases, it cannot integrate wholly into the culture and society. This makes the traditional African practitioners a vital part of their health care system. There are many reasons why the Western medical system has not been as effective in Africa as it has in more developed parts of the world. Hospitals and medical facilities are difficult for many Africans to get to. With vast areas of land and poor road and transportation systems, many native Africans have to travel immense distances on foot to reach help. Once they arrive they are often required to wait in line for up to 8 hours, especially in urban areas, as the lack of clinics and resources cause over-crowding. Patients are often not told the cause of their illness or much information about it all, so they have no way to prevent or prepare for it. The technology used is usually of poor quality, which impairs the quality of treatment. Modern medicine can also be too expensive for the average African to afford, making it difficult for them to receive proper care. Finally, Modern medicine removes native Africans from the culture and tradition and forces them into a setting that they are not comfortable with, away from their family and traditions which are of utmost importance to them. They do not get the proper spiritual healing that their culture seeks and traditional ideology requires.

However, there has been more interest expressed recently in the effects of some of the medicinal plants of Africa. "The pharmaceutical industry has come to consider traditional medicine as a source for identification of bio-active agents that can be used in the preparation of synthetic medicine." Pharmaceutical industries are looking into the medicinal effects of the most commonly and widely used plants to use in drugs. It's apparent that there are some things that can be learned from traditional African practice. In comparing the techniques of African healers and Western techniques, T. Adeoze Lambo, a Nigerian psychiatrist, stated in 1979, "At about three years ago, we made an evaluation, a programme of their work, and compared this with our own, and we discovered that actually they were scoring almost sixty percent success in their treatment of neurosis. And we were scoring forty percent-in fact, less than forty percent."

Effectiveness

Herbal medicines in Africa are generally not adequately researched, and are weakly regulated. There is a lack of the detailed documentation of the traditional knowledge, which is generally transferred orally. Several African medicinal plants have shown encouraging anti-trypanosomal effects but the research is only in the concept stage. A small proportion of ethnoveterinary medicine plants in South Africa have been researched for biological activity. Research identified favorable, possible future use of the *Hypoxis* species, (known locally as *inkomfe* or *African potato*) in both ATM and modern medicine. South African sangomas have been long and vocal advocates of a local traditional plant called *unwele* or *kankerbos* (*Sutherlandia frutescens*) claiming it assists in the treatment of HIV/AIDS, cancer and tuberculosis. Sufficient preclinical data on *Sutherlandia frutescens* yields plausible hypotheses that may account for the reputed efficacy.

Safety

A small proportion of ethnoveterinary medicine plants in South Africa have been researched for toxic effects. The possible adverse effects of South African traditional medicines are not well documented; there has been limited research into mutagenic properties and heavy metal contamination. Serious adverse effects, even death, can result from misidentification or misuse of healing plants. For example, various aloe plants are widely used in TAM, but some varieties, such as *Aloe globuligemma*, are toxic and can cause death. The potential for ATMs and pharmacokinetic interactions is unknown, especially interactions between traditional treatments and pharmaceutical antiretroviral drugs for HIV/AIDS. Herbal treatments are frequently used in Africa as a primary treatment for HIV/AIDS and for HIV-related issues. Collaboration with traditional healers has been recommended to determine what herbal remedies are used for HIV and to educate people supplying alternative treatments against unsafe practices. Given the demands of the local population on the use of ATMs, it has been suggested that South African medical schools should review their curriculum regarding traditional, complementary and alternate medicines.

Use of TM, using their antiviral efficiency instead of using specific antiretroviral drugs, is especially a risk with HIV. The hope to find a cure for Ebola or Marburg virus infections has been unsuccessful up to now. However, this problem is also true for modern medicine.

Boscia Senegalensis

Boscia senegalensis, or hanza, is a member of the family Capparaceae.

The plant originated from West Africa. Still a traditional food plant in Africa, this little-known fruit has potential to improve nutrition, boost food security, foster rural development and support sustainable landcare.

B. senegalensis is a perennial woody plant species of the *Boscia* genus in the caper (Capparaceae) family. This plant is classified as a dicot. Native to the Sahel region in Africa, this evergreen shrub can grow anywhere from 2 to 4 m (6 ft 7 in to 13 ft 1 in) in height under favourable conditions. The leaves of the plant are small and leathery, reaching 12 cm × 4 cm (4.7 in × 1.6 in). *B. senegalensis* produces fruits, clustered in small bunches, in the form of yellow spherical berries, up to 1.5 cm (0.59 in) in diameter. These fruits contain 1–4 seeds, which are a greenish hue when mature.

B. senegalensis is recognized as a potential solution to hunger and a buffer against famine in the Sahel region due to the variety of useful products it yields. It produces products for consumption, household needs, and medicinal and agricultural uses.

Other common names include: *aizen* (Mauritania), *mukheit* (Arabic), *hanza* (Hausa), *bere* (Bambara), *ngigili* (Fulani), and *mandiarha* (Berber). The fruits are also known as *dilo* (Hausa), *bokkhelli* (Arabic), *gigile* (Fulani).

History, Geography and Ethnography

B. senegalensis is a wild species, native to the Sahel region in Africa. It has not yet been domesticated. It currently grows in: Algeria, Benin, Burkina Faso, Cameroon, Central African Republic, Chad, Ghana, Guinea, Kenya, Mali, Mauritania, Niger, Nigeria, Senegal, Somalia, Sudan, and Togo.

Ethnobotanical indigenous knowledge contributes to the importance of this plant to the Hausa peoples of Niger and Fulani herders in West Africa. During the famine of 1984–1985, it was reported that *B. senegalensis* was the most widely consumed famine food in both Sudan and Darfur, relied on by over 94% of people in northern Darfur.

Growing Conditions

B. senegalensis grows in altitudes of 60–1,450 m (200–4,760 ft), in temperatures between 22–30 °C (72–86 °F) and with rainfall conditions of 100–500 mm (3.9–19.7 in) annually. It can be found growing in marginal soils: rocky, lateritic, clay stony hills, sand dunes, and sand-clay plains. These characteristics make it a highly resilient species, able to grow without expensive inputs even in the extremely hot and dry desert region of the Sahel. Herein lies its significance for poor farmers – in times of severe drought and famine, when many other crops have failed, *B. senegalensis* can still survive and provide useful products.

Other Farming Issues

Boscia senegalensis - Occurrence in field

B. senegalensis can benefit farmers because it keeps soil from laying bare and thus prevents soil erosion and degradation. It also buffers against wind, stabilizes sand dunes, offers shade to surrounding plants and cycles nutrients. In Niger, the trees are often cut or burned down by farmers in the dry season, in order to make space on the field for staple crops as millet or sorghum. However, due to the strong surviving character of the tree, it reappears after the first rains and continues growing as a small bush.

How Consumed and Uses

Fruits are ready for human consumption at the beginning of the rainy season, when most crops are just being planted, and there is little other food available. Fruits can be consumed raw and cooked. Raw fruits initially contain a sweet pulp that then dries out to a sugary solid, difficult to separate from seed. Fruits are often cooked prior to consumption. Juice can also be extracted and boiled down into a butter-like consistency that can be mixed with millet and milk to make cakes. In Sudan, the fruit is fermented into a beer.

The seeds of *B. senegalensis* are also important sources of nutrition, especially during times of famine. To gain access to the seeds, fruits are dried in the sun, pounded to remove the outer seed coat and soaked in water for several days, changing the water every day. The seed soaking process, also known as *debittering*, is essential to remove bitter and potentially toxic components. Seeds are usually cooked prior to consumption. Cooked seeds are texturally similar to a chickpea and can be used as a cereal substitute in stews, soups and porridges. Additionally, seeds can be re-dried and stored for later use or ground into a flour that can be used to make porridge. Roasted seeds can also serve as a substitute for coffee.

Hanza bread, cookies and cooked hanza, Zinder, Republic of Niger

Modern uses of *B. senegalensis* seeds are being developed in Niger Republic. They include cakes, cookies, bread, canned and popped seeds. These products from natural, wild *B. senegalensis* were recognised with the innovation award at an international food fair in Niamey, Niger, 2012.

Leaf extracts contain carbohydrate hydrolase enzymes that are useful for the production of cereal-based flour and for reducing the bulk of cereal porridges. Due to their proven biocidal activities, leaves are also added to granaries to protect cereals against pathogens. Leaves have many medicinal properties, notably anti-parasitic, fungicidal, anti-inflammatory and wound healing properties. Leaves, although not pleasant to taste, can be used as emergency forage for animals.

Young roots can be ground and boiled down into a thick, sweet porridge.

Wood can be used for home construction as well as for cooking fuel in times of dire need.

B. senegalensis contains natural coagulants that can be used to clarify water sources. Components of the plant (bark, twigs, leaves, fruits) can be added to a bucket of murky water, and the natural coagulants will cause clay and other particulates to compact and sink the bottom, allowing clear water to be obtained from the top.

Nutritional Information

Fruits are a significant source of carboydrates, as they contain 66.8% carbohydrates.

The seeds are sufficiently nutritious, although they do lack some essential nutrients, notably lysine and threonine. The seeds have significant levels of protein (25% of dry matter) and carbohydrates (60%). In these regards, seeds outperform local staple cereals such as sorghum and millet. Additionally, seeds are rich in zinc, iron, methionine, tryptophan, B-vitamins and linoleic acid (essential fatty acid). Seeds contain 3.6 times the World Health Organization (WHO) ideal level of tryptophan.

Leaves have high antioxidant capacity (nearly 1.5 times that of spinach) and are high in calcium, potassium, manganese and iron. The bioavailability of these compounds, however, is not very well known.

Economics

Leaves, seeds and fruits of *B. senegalensis* are traded in many small markets in the Sahel region. Some opportunities to add value are: roasting seeds to be sold as a coffee bean substitute, fermenting fruit into beer, processing fruit and seeds into prepared food, or processing leaves into medicinal applications. It can help raise incomes of the poor by protecting their stored cereals from pests and by substituting for other purchases from the market.

Gender Issues

Women in rural areas usually have the responsibility of gathering and preparing *B. senegalensis* for consumption. This process can create an extra work burden for women, however, their dominion over this process may result in increased access to this food source and thus contribute to improving their nutritional status.

Constraints to Wider Adoption

A major constraint to the wider adoption of *B. senegalensis* is the recalcitrant nature of its seeds. Seeds of this type are not well suited for ex-situ conservation, as they rapidly lose viability, and embryos are killed when seeds are dried. This creates a barrier to widespread growth, as it is difficult to propagate large numbers of plants for large-scale genetic selection and breeding. Other drawbacks to consumption include the issue of toxicity and the associated need to use scarce water resources and additional labour to leach out toxins during the debittering process.

Practical Information

One intervention with the potential to help poor farmers is the creation of cool temperature storage facilities – as *B. senegalensis* seeds can be stored for up to 2 months at 15 °C (59 °F).

It is imperative to spread knowledge of the wide range of benefits that *B. senegalensis* provides, in order to encourage small farmers to plant it. New plantings would offer increased protection to the soil as well as provide food and other resources in times of famine.

It is recommended that the techniques of grafting and generating hybrids (wide-crosses) with related species be explored, as both techniques have the potential to increase harvests and/or improve the fruits. Promising preliminary research is being conducted using in vitro tissue culture technologies to propagate *B. senegalensis*. Additionally, direct seedling trials are recommended and being advanced by the Eden Foundation.

Leonotis Leonurus

Leonotis leonurus, also known as lion's tail and wild dagga, is a plant species in the Lamiaceae (mint) family. The plant is a broadleaf evergreen large shrub native to South Africa and southern Africa, where it is very common. It is known for its medicinal and mild psychoactive properties. The main psychoactive component of *Leonotis leonurus* is leonurine. The name 'wild dagga' links it closely to cannabis as 'dagga' derived from the Khoikhoi 'dachab' is an indigenous South African name for cannabis species.

Close-up: *Leonotis leonurus* flower.

Description

The shrub grows 3 to 6 ft (1 to 2 m) tall by 1.5 to 3.5 feet (0.46 to 1.07 m) wide. The medium-dark green 2–4 inches (5.1–10.2 cm) long leaves are aromatic when crushed. The plant has tubular orange flowers in tiered whorls, typical to the mint family, that encircle the square stems. They rise above the foliage mass during the summer season, with flowering continuing into winter in warmer climates.

Variation in Flower Color

A white variety (known colloquially as 'Alba') and a yellow variety also exist.

Ecology

In its native habitats *Leonotis leonurus* attracts nectivorous birds (mainly sunbirds), as well as various insects such as butterflies. The flowers' mainly orange to orange-red colour and tubular shape are indicative of its co-evolution with African sunbirds, which have curved bills suited to feeding from tubular flowers.

Cultivation

Leonotis leonurus is cultivated as an ornamental plant for its copious orange blossom spikes and accent or screening qualities for use in gardens and parks. It is a moderate drought tolerant plant, and a nectar source for birds and butterflies in landscape settings.

Lion's tail can especially be found in other subtropical and Mediterranean climate regions beyond South Africa such as California, Hawaii, and Australia where it has naturalized in areas. In cooler climates it is used as an annual and winter conservatory plant.

Pharmacology and Toxicology

The main active component leonurine has both antioxidant and cardioprotective properties and has shown to significantly improve myocardial function.

One experimental animal study suggests that the aqueous leaf extract of *Leonotis leonurus* possesses antinociceptive, antiinflammatory, and hypoglycemic properties.

An animal study in rats indicated that in high doses, lion's tail has significant toxicological adverse effects on organs, red blood cells, white blood cells and other important bodily functions. Acute toxicity tests in animals caused death for those receiving 3200 mg/kg dose. At 1600 mg/kg extract led to changes in red blood cells, hemoglobin concentration, mean corpuscular volume, platelets, and white blood cells.

Traditional Uses

The infusions made from flowers and seeds, leaves or stems are widely used to treat tuberculosis, jaundice, muscle cramps, high blood pressure, diabetes, viral hepatitis, dysentery, and diarrhoea. The leaves, roots and bark are used as an emetic for snakebites, bee and scorpion stings. The fresh stem juice is used as an infusion drunk for 'blood impurity' in some places of South Africa.

Recreational Uses

The dried leaves and flowers have a mild calming effect when smoked. In some users, the effects have been noted to be similar to the cannabinoid THC found in *Cannabis*, except that it has a much less potent high. It has also been reported to cause mild euphoria, visual changes, dizziness, nausea, sweating, sedation and lightheadedness.

It is sometimes used as a *Cannabis* substitute by recreational users as an alternative to illegal psychoactive plants.

Legal Status

Latvia

Leonotis leonurus is illegal in Latvia since November 2009. It is Schedule 1 drug. Possession of quantities up to 1 gram are fined up to 280 euro. Possession and distribution of larger quantities can be punished with up to 15 years in prison.

Poland

Leonotis leonurus was banned in Poland in March 2009. Possession and distribution lead to a criminal charge.

United States

Leonotis leonorus is not currently scheduled under federal law in the United States.

Agathosma Betulina

Agathosma betulina (previously *Barosma betulina*) is a flowering plant in the family Rutaceae, native to the lower elevation mountains of western South Africa, where it occurs near streams in fynbos habitats.

Growth

It is an evergreen shrub growing to 2 m tall. The leaves are opposite, rounded, about 20 mm long and broad. The flowers are white or pale pink, with five petals; the fruit is a five-parted capsule which splits open to release the seeds.

Etymology

It is known by the common name round leaf buchu and has historically been used as a flavoring agent and an herbal remedy. The very similar plant *Agathosma crenulata* (previously *Barosma crenulata*) is known as the oval leaf buchu, and has been used for the same purposes.

Uses

Wild plants of this species are still plentiful but are being harvested faster than they can reproduce. The threat of their becoming scarce has led to efforts to cultivate them. The essential oils and extracts of the leaves are used as flavoring for teas, candy, and a liquor known as buchu brandy in South Africa. The two primary chemical constituents of the oils of *A. betulina* are isomenthone and diosphenol. The extract is said to taste like blackcurrant.

Folk Medicine

The plant has been used by the indigenous people of South Africa to as a folk remedy for various disorders. Dutch settlers in early times used Agathosma betulina commonly called buchu to make a brandy tincture. The tincture is still used today.

Harpagophytum

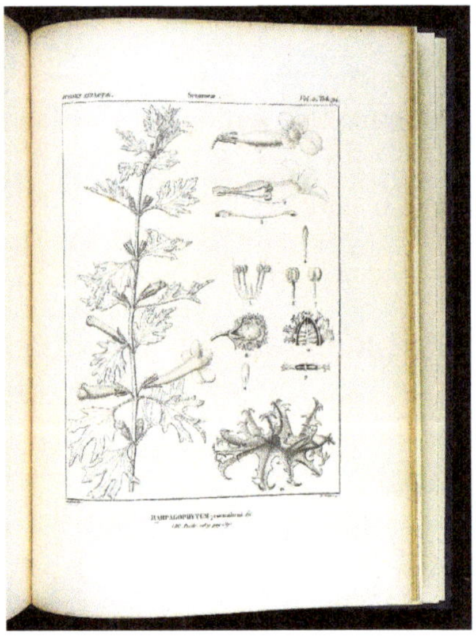

Plate from "Icones selectae plantarum", vol. 5: t. 94 (1846)

Harpagophytum procumbens, also called grapple plant, wood spider and most commonly devil's claw, is a genus of plants in the sesame family, native to southern Africa. Plants of the genus owe their common name "devil's claw" to the peculiar appearance of their hooked fruit. Several species of North American plants in genus *Proboscidea* and certain species of *Pisonia* are however also known by this name. Devil's claw's tuberous roots are used in folk medicine to reduce pain.

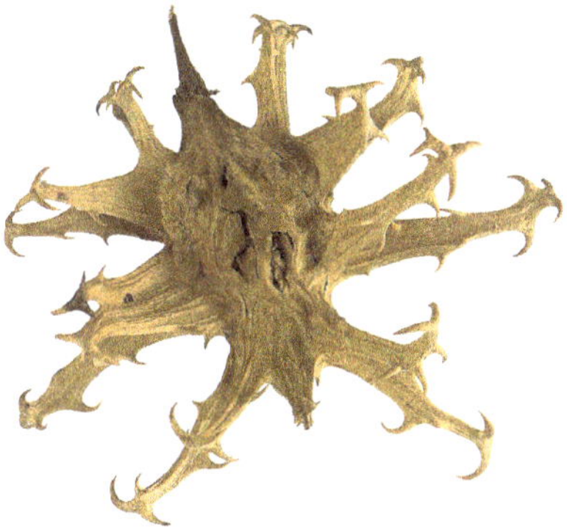

Dry fruit of *H. procumbens* - MHNT

Range

H. procumbens is mainly found in the eastern and south eastern parts of Namibia, Southern Botswana and the Kalahari region of the Northern Cape, South Africa. *H. zeyheri* is found in the northern parts of Namibia (Ovamboland) and southern Angola.

Folk Medicine and Research

The ethnobotanical use of devil's claw originated in southern Africa. *H. procumbens* is one of the floral emblems of Botswana where it is thought to be useful in treating a variety of pain conditions.

Preparations of the plant or its extracts, such as harpagoside, are presumed to have uses in folk medicine and phytotherapy as an anti-inflammatory herbal drug or dietary supplement. Although there is no accepted clinical evidence of its efficacy and bioavailability, limited effects were noted for treating lower back pain and osteoarthritis.

A Cochrane review of clinical research noted that devil's claw seems to reduce low back pain more than placebo, although evidence was of moderate quality at best.

Adverse Reactions

Side effects and drug interactions with nonsteroidal anti-inflammatory drugs may occur, and caution exists for a variety of conditions, such as pregnancy and cardiovascular disorders. Devil's claw may cause diarrhea and may interfere with the action of ticlopidine and warfarin.

Ecology and Management

H. procumbens inhabits deep, sandy soils, and occurs in areas with low annual rainfall (150–300 mm/year). It is a perennial, tuberous plant with annually produced creeping stems. The aboveground stems emerge after the first rains and die back during droughts or after frosts. The stems grow from a persistent primary tuber and several secondary tubers (the harvested organs) grow from the primary tuber at the end of fleshy roots. The plant gets its scientific and common names from the hooked spines of its woody capsules. The mature fruit opens slowly so that, in a given year, only 20-25% of its seeds may establish soil contact. Seeds have a high degree of dormancy. They have a low respiration rate and may remain viable in the seed bank for more than 20 years.

The sustainability of the trade in devil's claw has been questioned for several years. The governments of the each of the countries in which it occurs (range states; Namibia, Botswana, and South Africa) have developed policies and regulations to protect the species, to determine a sustainable harvest, and to provide for continued livelihoods for the harvesters. At various times, the species has been proposed for protection by the Convention of International Trade in Endangered Species (CITES). However, the range states have implemented measures to manage the trade sustainably and the proposal to protect the species by CITES was withdrawn.

Various studies have examined the biological and ecological requirements of harvested and unharvested populations. Several early short-term studies in Botswana examined the ecological requirements of the species. Other, somewhat more recent studies inventoried the resource and examined sustainable harvesting methods.

The ecological requirements of a harvested species must be known to manage the harvest of the species to ensure it is sustainable. Stewart & Cole (2005) examined the complex economic, social, and cultural factors involved in the harvest of the species. Stewart (2009) examined population structure, density, growth, mortality, and seed and fruit production in harvested and unharvested populations in the Kalahari savannas of South Africa. Plant density and population structure differed significantly between overgrazed and grass-dominated areas, suggesting that the differences may be due to competition for scarce water and nutrients. Experimental removal of secondary tubers (harvest) was not a significant factor for mortality in any of the harvested size classes. Harvest also did not affect growth, although plants in the medium size class grew more during the study period in both the harvested and unharvested populations. Fruit production was highly variable, and mature fruits were produced only under favorable conditions. Under the conditions of this experimental harvest, the species appears to be resilient to harvest, with plants subjected to harvest surviving as well as unharvested plants. However, due to the spatially variable nature of its habitat and the plasticity of the plants themselves, harvesting data from actual harvested areas from a large number of plants is required to better understand the life history of the species.

Pelargonium Cucullatum

Pelargonium cucullatum. White-flowered variety. The flower on the left is younger and still has its stamens, though the pistil has not emerged properly yet. Its maturer neighbour on the right has dropped its anthers and deployed its stigma.

Pelargonium cucullatum (also called wildemalva, tree pelargonium or hooded-leaf pelargonium) is a species of plant in the Geraniaceae family, that is indigenous to the south-western Cape of South Africa. It produces masses of pink and purple flowers in the summer and has been used to produce a great number of modern Pelargonium hybrids.

This vigorous, fast-growing shrub can reach heights of over a meter. The leaves grow upwards, forming circular bowls with jagged, red-tipped edges. The sweetly scented flowers are purple or pink and appear for several months over the summer. During this time, the wild plants stand out in the indigenous fynbos in places like Table Mountain.

There are three subspecies, *Pelargonium cucullatum* subsp. *cucullatum*, *Pelargonium cucullatum* subsp. *strigifolium* and *Pelargonium cucullatum* subsp. *tabulare*

Subspecies *tabulare* is found in the Cape Peninsula.

Pelargonium cucullatum is becoming increasingly popular as a garden plant, especially in the Western Cape, South Africa. It prefers a sunny position and once established is very tough and waterwise. It can be propagated by cuttings and by seed. Like many geraniums, this plant's leaves give off a pleasant herbal scent when crushed and they have a wide variety of uses in traditional medicine.

References

- Guneratne, Arjun (2009). Culture and the Environment in the Himalaya. Routledge contemporary South Asia series, #24. New York: Routledge. pp. 84–85. ISBN 978-0-415-77883-1
- Sekar, S. (2007). "Traditional alcoholic beverages from Ayurveda and their role on human health". Indian Journal of Traditional Knowledge. 6 (1): 144–149. Archived from the original on 18 December 2014
- Wujastyk, Dominik (2003). The Roots of Ayurveda: Selections from Sanskrit Medical Writings (3 ed.). London etc.: Penguin Books. ISBN 0-140-44824-1
- Wujastyk, Dominik (2005). "Policy Formation and Debate Concerning the Government Regulation of Ayurveda in Great Britain in the 21st Century". Asian Medicine. 1: 162–184. doi:10.1163/157342105777996719
- Sujatha, V (July 2011). "What could 'integrative' medicine mean? Social science perspectives on contemporary Ayurveda". Journal of Ayurveda and Integrative Medicine. 2 (3): 115–23. PMC 3193682 . PMID 22022153. doi:10.4103/0975-9476.85549
- Frawley, David; Ranade, Subhash (2001). Ayurveda, Nature's Medicine. Lotus Press. p. 11. ISBN 9780914955955. Retrieved 6 January 2015
- Padalia, Rajendra C.; Verma, Ram S. (2011). "Comparative volatile oil composition of four Ocimum species from northern India". Natural Product Research. 25 (6): 569–575. PMID 21409717. doi:10.1080/14786419.2010.482936
- Quack, Johannes (2011). Disenchanting India: Organized Rationalism and Criticism of Religion in India. Oxford University Press. pp. 213, 3. ISBN 9780199812608
- Ernst, E. (2002). "Heavy metals in traditional Indian remedies". European Journal of Clinical Pharmacology. 57 (12): 891–896. ISSN 0031-6970. PMID 11936709. doi:10.1007/s00228-001-0400-y
- "Civic Heraldry of Namibia". Heraldry of the World. Ralf Hartemink. Archived from the original on 27 December 2007. Retrieved 31 March 2013
- Wujastyk, Dominik (2003). "Indian Medicine". In Flood, Gavin. The Blackwell Companion to Hinduism. Oxford: Blackwell. p. 394. ISBN 978-1405132510

- "A Bangladeshi Wedding Journal – Gaye Holud: Pre-Wedding Ceremony". The Daily Star. November 11, 2014. Retrieved February 22, 2017
- Zysk, Kenneth G. (1998). Asceticism and Healing in Ancient {India}: Medicine in the {Buddhist} Monastery. Delhi: Motilal Banarsidass. ISBN 81-208-1507-6
- Bhattacharya, A.; Chatterjee, A.; Ghosal, S.; Bhattacharya, S. K. (1999). "Antioxidant activity of active tannoid principles of Emblica officinalis (amla)". Indian journal of experimental biology. 37 (7): 676–680. PMID 10522157
- Priyadarsini, KI (2014). "The chemistry of curcumin: from extraction to therapeutic agent". Molecules. 19 (12): 20091–112. PMID 25470276. doi:10.3390/molecules191220091
- Brett, F. H. (1840). A practical essay on some of the principal surgical diseases of India. Calcutta, India: W. Thacker & Company. p. 458. Retrieved 6 January 2015
- Bausell, R. Barker (2007). Snake Oil Science: The Truth About Complementary and Alternative Medicine. Oxford University Press. p. 259. ISBN 9780195383423
- Tayyem RF, Heath DD, Al-Delaimy WK, Rock CL (2006). "Curcumin content of turmeric and curry powders". Nutr Cancer. 55 (2): 126–131. PMID 17044766. doi:10.1207/s15327914nc5502_2
- "Legal Status of Traditional Medicine and Complementary/Alternative Medicine: A Worldwide Review". World Health Organization (WHO). p. 132. Retrieved June 24, 2014

4

Herbalism: An Integrated Study

Humans have been using plants for medicinal practices for thousands of years. Herbal medicine is considered to be a form of alternative medicine in today's time. Decoction and zoopharmacognosy are some of the techniques used to make herbal medicine. This chapter will provide an integrated understanding of herbalism.

Herbalism

Herbalism (also herbal medicine or phytotherapy) is the study of botany and use of plants intended for medicinal purposes or for supplementing a diet. Plants have been the basis for medical treatments through much of human history, and such traditional medicine is still widely practiced today. Modern medicine recognizes herbalism as a form of alternative medicine, as the practice of herbalism is not strictly based on evidence gathered using the scientific method. Modern medicine makes use of many plant-derived compounds as the basis for evidence-based pharmaceutical drugs. Although phytotherapy may apply modern standards of effectiveness testing to herbs and medicines derived from natural sources, few high-quality clinical trials and standards for purity or dosage exist. The scope of herbal medicine is sometimes extended to include fungal and bee products, as well as minerals, shells and certain animal parts.

Synonyms

As terms referring to medicinal or dietary practices of using botanical products, *herbalism*, *herbal medicine* or *phytotherapy* are used interchangeably in many countries, including Canada, Norway, the United Kingdom, other countries in Europe and South America, South Africa, and the United States.

General practices include ancient methods of traditional Chinese medicine and Ayurveda. Practitioners of herbalism or phytotherapy are referred to as herbalists or phytotherapists. Products used in herbalism may be called herbal medicines, botanicals, natural health products, herbal remedies, herbal supplements, or phytotherapies.

History

Archaeological evidence indicates that the use of medicinal plants dates back to the Paleolithic age, approximately 60,000 years ago. Written evidence of herbal remedies dates back over 5,000 years, to the Sumerians, who compiled lists of plants. A number of ancient cultures wrote about plants and their medical uses in books called *herbals*. In ancient Egypt, herbs are mentioned in Egyptian medical papyri, depicted in tomb illustrations, or on rare occasions found in medical jars containing trace amounts of herbs. Among the oldest, lengthiest, and most important medical papyri of ancient Egypt, the Ebers Papyrus dates from about 1550 BC, and covers more than 700 drugs, mainly of plant origin. The earliest known Greek herbals come from Theophrastus of Eresos who in the 4th c. B.C. wrote in Greek *Historia Plantarum*, from Diocles of Carystus who wrote during the 3rd century B.C, and from Krateuas who wrote in the 1st century B.C. Only a few fragments of these works have survived intact, but from what remains scholars have noted a large amount of overlap with the Egyptian herbals. Seeds likely used for herbalism have been found in archaeological sites of Bronze Age China dating from the Shang Dynasty (c. 1600 BC–c. 1046 BC). Over a hundred of the 224 drugs mentioned in the *Huangdi Neijing*, an early Chinese medical text, are herbs. Herbs also commonly featured in the medicine of ancient India, where the principal treatment for diseases was diet. *De Materia Medica*, originally written in Greek by Pedanius Dioscorides (c. 40 – 90 AD) of Anazarbus, Cilicia, a Greek physician, pharmacologist and botanist, is a particularly important example of herbal writing; it dominated for some 1500 years until the 1600s.

Modern Herbal Medicine

The World Health Organization (WHO) estimates that 80 percent of the population of some Asian and African countries presently use herbal medicine for some aspect of primary health care. Pharmaceuticals are prohibitively expensive for most of the world's population, half of whom lived on less than $2 U.S. per day in 2002. In comparison, herbal medicines can be grown from seed or gathered from nature for little or no cost.

Many of the pharmaceuticals currently available to physicians have a long history of use as herbal remedies, including opium, aspirin, digitalis, and quinine. According to the World Health Organization, approximately 25% of modern drugs used in the United States have been derived from plants. At least 7,000 medical compounds in the modern pharmacopoeia are derived from plants. Among the 120 active compounds currently isolated from the higher plants and widely used in modern medicine today, 80% show a positive correlation between their modern therapeutic use and the traditional use of the plants from which they are derived.

Clinical Tests

In a 2010 global survey of the most common 1000 plant-derived compounds, only 156 had clinical trials published. Preclinical studies (cell culture and animal studies) were reported for about one-half of the plant products, while 12% of the plants, although available in the Western market, had "no substantial studies" of their properties. Strong evidence was found that 5 were toxic or allergenic, so that their use ought to be discouraged or forbidden. Nine plants with evidence of therapeutic effect included *Althaea officinalis, Calendula officinalis, Centella asiatica, Echinacea purpurea, Passiflora incarnata, Punica granatum, Vaccinium macrocarpon, Vaccinium myrtillus*, and *Valeriana officinalis.*

In 2015, the Australian Government's Department of Health published the results of a review of alternative therapies that sought to determine if any were suitable for being covered by health insurance; Herbalism was one of 17 topics evaluated for which no clear evidence of effectiveness was found.

According to Cancer Research UK, "there is currently no strong evidence from studies in people that herbal remedies can treat, prevent or cure cancer".

Establishing guidelines to assess safety and efficacy of herbal products, the European Medicines Agency provides criteria for evaluating and grading the quality of clinical research in preparing monographs about herbal products.

In the United States, the National Center for Complementary and Integrative Health of the National Institutes of Health funds clinical trials on herbal compounds, provides fact sheets evaluating the safety, potential effectiveness and side effects of many plant sources, and maintains a registry of clinical research conducted on herbal products.

Prevalence of Use

The use of herbal remedies is more prevalent in patients with chronic diseases such as cancer, diabetes, asthma and end-stage renal disease. Multiple factors such as gender, age, ethnicity, education and social class are also shown to have association with prevalence of herbal remedies use.

A survey released in May 2004 by the National Center for Complementary and Integrative Health focused on who used complementary and alternative medicines (CAM), what was used, and why it was used. The survey was limited to adults, aged 18 years and over during 2002, living in the United States. According to this survey, herbal therapy, or use of natural products other than vitamins and minerals, was the most commonly used CAM therapy (18.9%) when all use of prayer was excluded.

Herbal remedies are very common in Europe. In Germany, herbal medications are dispensed by apothecaries (e.g., Apotheke). Prescription drugs are sold alongside essential oils, herbal extracts, or herbal teas. Herbal remedies are seen by some as a treatment to be preferred to pure medical compounds that have been industrially produced.

In India the herbal remedy is so popular that the government of India has created a separate department—AYUSH—under the Ministry of Health & Family Welfare. The National Medicinal Plants Board was also established in 2000 by the Indian government in order to deal with the herbal medical system.

Herbal Preparations

There are many forms in which herbs can be administered, the most common of which is in the form of a liquid that is drunk by the patient—either an herbal tea or a (possibly diluted) plant extract.

Several methods of standardization may be determining the amount of herbs used. One is the ratio of raw materials to solvent. However different specimens of even the same plant species may vary in chemical content. For this reason, thin layer chromatography is sometimes used by growers to assess the content of their products before use. Another method is standardization on a signal chemical.

Herbal teas, or tisanes, are the resultant liquid of extracting herbs into water, though they are made in a few different ways. Infusions are hot water extracts of herbs, such as chamomile or mint, through steeping. Decoctions are the long-term boiled extracts, usually of harder substances like roots or bark. Maceration is the old infusion of plants with high mucilage-content, such as sage, thyme, etc. To make macerates, plants are chopped and added to cold water. They are then left to stand for 7 to 12 hours (depending on herb used). For most macerates 10 hours is used.

Leaves of *Eucalyptus olida* being packed into a steam distillation unit to gather its essential oil.

Tinctures are alcoholic extracts of herbs, which are generally stronger than herbal teas. Tinctures are usually obtained by combining 100% pure ethanol (or a mixture of 100% ethanol with water) with the herb. A completed tincture has an ethanol percentage of at least 25% (sometimes up to 90%). Herbal wine and elixirs are alcoholic extract of herbs, usually with an ethanol percentage of 12–38%. Extracts include liquid extracts, dry extracts, and nebulisates. Liquid extracts are liquids with a lower ethanol percentage than tinctures. They are usually made by vacuum distilling tinctures. Dry extracts are extracts of plant material that are evaporated into a dry mass. They can then be further refined to a capsule or tablet.

The exact composition of an herbal product is influenced by the method of extraction. A tea will be rich in polar components because water is a polar solvent. Oil on the other hand is a non-polar solvent and it will absorb non-polar compounds. Alcohol lies somewhere in between.

A herb shop in the souk of Marrakesh, Morocco

Many herbs are applied topically to the skin in a variety of forms. Essential oil extracts can be applied to the skin, usually diluted in a carrier oil. Many essential oils can burn the skin or are simply too high dose used straight; diluting them in olive oil or another food grade oil such as almond oil can allow these to be used safely as a topical. Salves, oils, balms, creams and lotions are other forms of topical delivery mechanisms. Most topical applications are oil extractions of herbs.

Taking a food grade oil and soaking herbs in it for anywhere from weeks to months allows certain phytochemicals to be extracted into the oil. This oil can then be made into salves, creams, lotions, or simply used as an oil for topical application. Many massage oils, antibacterial salves, and wound healing compounds are made this way.

Inhalation, as in aromatherapy, can be used as a treatment.

Safety

Datura stramonium has been used in Ayurveda for various treatments, but contains alkaloids, such as atropine and scopolamine, which may cause severe toxicity.

A number of herbs are thought to be likely to cause adverse effects. Furthermore, "adulteration, inappropriate formulation, or lack of understanding of plant and drug interactions have led to adverse reactions that are sometimes life threatening or lethal." Proper double-blind clinical trials are needed to determine the safety and efficacy of each plant before they can be recommended for medical use. Although many consumers believe that herbal medicines are safe because they are "natural", herbal medicines and synthetic drugs may interact, causing toxicity to the patient. Herbal remedies can also be dangerously contaminated, and herbal medicines without established efficacy, may unknowingly be used to replace medicines that do have corroborated efficacy.

Standardization of purity and dosage is not mandated in the United States, but even products made to the same specification may differ as a result of biochemical variations within a species of plant. Plants have chemical defense mechanisms against predators that can have adverse or lethal effects on humans. Examples of highly toxic herbs include poison hemlock and nightshade. They are not marketed to the public as herbs, because the risks are well known, partly due to a long and colorful history in Europe, associated with "sorcery", "magic" and intrigue. Although not frequent, adverse reactions have been reported for herbs in widespread use. On occasion serious untoward outcomes have been linked to herb consumption. A case of major potassium depletion has been attributed to chronic licorice ingestion., and consequently professional herbalists avoid the use of licorice where they recognize that this may be a risk. Black cohosh has been implicated in a case of liver failure. Few studies are available on the safety of herbs for pregnant women, and one study found that use of complementary and alternative medicines are associated with a 30% lower ongoing pregnancy and live birth rate during fertility treatment. Examples of herbal treatments with likely cause-effect relationships with adverse events include aconite, which is often a legally restricted herb, ayurvedic remedies, broom, chaparral, Chinese herb mixtures, comfrey, herbs containing certain flavonoids, germander, guar gum, liquorice root, and pennyroyal. Examples of herbs where a high degree of confidence of a risk long term adverse effects can be asserted include ginseng, which is unpopular among herbalists for this reason, the endangered herb goldenseal, milk thistle, senna, against which herbalists generally advise and rarely use, aloe vera juice, buckthorn bark and berry, cascara sagrada bark, saw palmetto, valerian, kava, which is banned in the European Union, St. John's wort, Khat, Betel nut, the restricted herb Ephedra, and Guarana.

There is also concern with respect to the numerous well-established interactions of herbs and drugs. In consultation with a physician, usage of herbal remedies should be clarified, as some herbal remedies have the potential to cause adverse drug interactions when used in combination with various prescription and over-the-counter pharmaceuticals, just as a patient should inform a herbalist of their consumption of orthodox prescription and other medication.

For example, dangerously low blood pressure may result from the combination of an herbal remedy that lowers blood pressure together with prescription medicine that has the same effect. Some herbs may amplify the effects of anticoagulants. Certain herbs as well as common fruit interfere with cytochrome P450, an enzyme critical to much drug metabolism.

Labeling Accuracy

A 2013 study found that one-third of herbal supplements sampled contained no trace of the herb listed on the label. The study found products adulterated with contaminants or fillers not listed on the label, including potential allergens such as soy, wheat, or black walnut. One bottle labeled as St. John's Wort was found to actually contain *Alexandrian senna*, a laxative.

Researchers at the University of Adelaide found in 2014 that almost 20 per cent of herbal remedies surveyed were not registered with the Therapeutic Goods Administration, despite this being a condition for their sale. They also found that nearly 60 per cent of products surveyed had ingredients that did not match what was on the label. Out of 121 products, only 15 had ingredients that matched their TGA listing and packaging.

In 2015, the New York Attorney General issued cease and desist letters to four major U.S. retailers (GNC, Target, Walgreens, and Walmart) who were accused of selling herbal supplements that were mislabeled and potentially dangerous. Twenty-four products were tested by DNA barcoding as part of the investigation, with all but five containing DNA that did not match the product labels.

Practitioners of Herbalism

A herbalist gathers the flower heads of *Arnica montana*.

A herbalist is:

1. A person whose life is dedicated to the economic or medicinal uses of plants.
2. One skilled in the harvesting and collection of medicinal plants.
3. Traditional Chinese herbalist: one who is trained or skilled in the dispensing of herbal prescriptions; traditional Chinese herb doctor. Similarly, traditional Ayurvedic herbalist: one who is trained or skilled in the dispensing of herbal prescriptions in the Ayurvedic tradition.
4. One trained or skilled in the therapeutic use of medicinal plants.
5. One who is skilled in the preparation/manufacture of dried and/or liquid herbal products who possesses a pharmacognostic, formulary and/or clinical understanding of the products being prepared/manufactured.

Herbalists must learn many skills, including the wildcrafting or cultivation of herbs, diagnosis and treatment of conditions or dispensing herbal medication, and preparations of herbal medications. Education of herbalists varies considerably in different areas of the world. Lay herbalists and traditional indigenous medicine people generally rely upon apprenticeship and recognition from their communities in lieu of formal schooling.

In some countries formalized training and minimum education standards exist, although these are not necessarily uniform within or between countries. For example, in Australia the currently self-regulated status of the profession (as of April 2008) results in different associations setting different educational standards, and subsequently recognising an educational institution or course of training. The National Herbalists Association of Australia is generally recognised as having the most rigorous professional standard within Australia. In the United Kingdom, the training of medical herbalists is done by state funded Universities. For example, Bachelor of Science degrees in herbal medicine are offered at Universities such as University of East London, Middlesex University, University of Central Lancashire, University of Westminster, University of Lincoln and Napier University in Edinburgh at the present.

Government Regulations

The World Health Organization (WHO), the specialized agency of the United Nations (UN) that is concerned with international public health, published *Quality control methods for medicinal plant materials* in 1998 in order to support WHO Member States in establishing quality standards and specifications for herbal materials, within the overall context of quality assurance and control of herbal medicines.

In the European Union (EU), herbal medicines are regulated under the Committee on Herbal Medicinal Products.

In the United States, herbal remedies are regulated dietary supplements by the Food and Drug Administration (FDA) under current good manufacturing practice (cGMP) policy for dietary supplements. Manufacturers of products falling into this category are not required to prove the safety or efficacy of their product so long as they do not make 'medical' claims or imply uses other than as a 'dietary supplement', though the FDA may withdraw a product from sale should it prove harmful.

Canadian regulations are described by the Natural and Non-prescription Health Products Directorate which requires an eight-digit Natural Product Number or Homeopathic Medicine Number on the label of licensed herbal medicines or dietary supplements.

Some herbs, such as cannabis and coca, are outright banned in most countries though coca is legal in most of the South American countries where it is grown. The *Cannabis* plant is used as an herbal medicine, and as such is legal in some parts of the world. Since 2004, the sales of ephedra as a dietary supplement is prohibited in the United States by the FDA, and subject to Schedule III restrictions in the United Kingdom.

Scientific Criticism

Herbalism has been criticized as a potential "minefield" of unreliable product quality, safety hazards, and potential for misleading health advice. Globally, there are no standards across various herbal products to authenticate their contents, safety or efficacy, and there is generally an absence of high-quality scientific research on product composition or effectiveness for anti-disease activity.

Unethical practices by some herbalists and manufacturers, which may include false advertising about health benefits on product labels or literature, and contamination or use of fillers during product preparation, may erode consumer confidence about services and products.

Phytotherapy

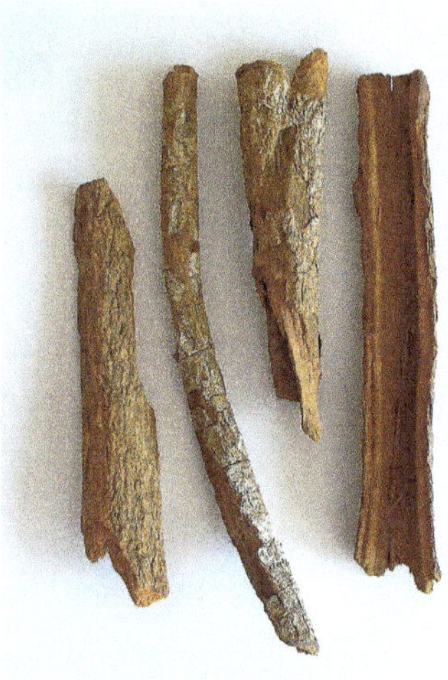

An example of a 'phytotherapeutic' compound: the bark of the cinchona tree contains quinine, which today is a widely prescribed treatment for malaria, especially in countries that cannot afford to purchase more expensive antimalarial drugs

Closely related to herbalism, phytotherapy is the intended medical use of plants and plant extracts for therapeutic purposes. A possible differentiation with herbalism is that phytotherapy may require constituents in the plant extract be standardized by adhering to a minimum content of one or several active compounds in the therapeutic product.

Modern phytotherapy may use conventional methods to assess herbal drug quality, but more typically relies on modern processes like high-performance liquid chromatography (HPLC), gas chromatography, ultraviolet/visible spectrophotometry or atomic absorption spectroscopy to identify species, measure bacteriological contamination, assess potency, and create Certificates of Analysis for the material.

Phytotherapy is distinct from homeopathy and anthroposophic medicine, and avoids mixing plant and synthetic bioactive substances. Phytotherapy is regarded by some as traditional medicine.

Traditional Herbal Medicine Systems

Ready to drink macerated medicinal liquor with goji berry, tokay gecko, and ginseng, for sale at a traditional medicine market in Xi'an, China.

Africa

Up to 80% of the population in Africa uses traditional medicine as primary health care.

Americas

Native Americans medicinally used about 2,500 of the approximately 20,000 plant species that are native to North America.

China

Some researchers trained in both western and traditional Chinese medicine have attempted to deconstruct ancient medical texts in the light of modern science. One idea is that the yin-yang balance, at least with regard to herbs, corresponds to the pro-oxidant and anti-oxidant balance. This interpretation is supported by several investigations of the ORAC ratings of various yin and yang herbs.

India

In India, Ayurvedic medicine has quite complex formulas with 30 or more ingredients, including a sizable number of ingredients that have undergone "alchemical processing", chosen to balance "Vata", "Pitta" or "Kapha".

In Ladakh, Lahul-Spiti and Tibet, the Tibetan Medical System is prevalent, also called the 'Amichi Medical System'. Over 337 species of medicinal plants have been documented by C.P. Kala. Those are used by Amchis, the practitioners of this medical system.

In Tamil Nadu, Tamils have their own medicinal system now popularly called Siddha medicine. The Siddha system is entirely in the Tamil language. It contains roughly 300,000 verses covering diverse aspects of medicine. This work includes herbal, mineral and metallic compositions used as medicine. Ayurveda is in Sanskrit, but Sanskrit was not generally used as a mother tongue and hence its medicines are mostly taken from Siddha and other local traditions.

Indonesia

Different types of Indonesian jamu herbal medicines held in bottles.

In Indonesia, especially among the Javanese, the jamu traditional herbal medicine is an age old tradition preserved for centuries. Jamu is thought to have originated in the Mataram Kingdom era, some 1300 years ago. The bas-reliefs on Borobudur depicts the image of people grinding herbs with stone mortar and pestle, a drink seller, a physician and masseuse treating their clients. All of these scenes might be interpreted as a traditional herbal medicine and health-related treatments in ancient Java. The Madhawapura inscription from Majapahit period mentioned a specific profession of herbs mixer and combiner (herbalist), called *Acaraki*. The medicine book from Mataram dated from circa 1700 contains 3,000 entries of jamu herbal recipes, while Javanese classical literature Serat Centhini (1814) describes some jamu herbal concoction recipes.

Though highly possible influenced by Indian Ayurveda system, Indonesia is a vast archipelago with numerous indigenous plants not to be found in India, which include plants similar to Australia beyond the Wallace Line. Indonesians might experimented and figure out the medicinal uses of these native herbal plants. Jamu may vary from region to region, and often not written down, especially in remote areas of the country. Although primarily herbal, materials acquired from animals, such as honey, royal jelly, milk and *ayam kampung* eggs are also often used in jamu.

Herbal Philosophy and Spiritual Practices

According to Eisenburg: "The Chinese and Western medical models are like two frames of reference in which identical phenomena are studied. Neither frame of reference provides an unobstructed view of health and illness. Each is incomplete and in need of refinement." Specifically, the traditional Chinese medical model could effect change on the recognized and expected phenomena of detachment to patients unique to the clinical relationships between patient and physician of the Western school of medicine.

Four approaches to the use of plants as medicine include:

1. The magical/shamanic—Almost all societies, with the exception of cultures influenced by Western-style industrialization, recognize this kind of use. The practitioner is regarded as endowed with gifts or powers that allow him/her to use herbs in a way that is hidden from the average person, and the herbs are said to affect the spirit or soul of the person.
2. The energetic—This approach includes the major systems of Traditional Chinese Medicine, Ayurveda, and Unani. Herbs are regarded as having actions in terms of their energies and affecting the energies of the body. The practitioner may have extensive training, and ideally be sensitive to energy, but need not have supernatural powers.
3. The functional dynamic—This approach was used by early physiomedical practitioners, whose doctrine forms the basis of contemporary practice in the UK. Herbs have a functional action, which is not necessarily linked to a physical compound, although often to a physiological function, but there is no explicit recourse to concepts involving energy.
4. The chemical—Modern practitioners – called Phytotherapists – attempt to explain herb actions in terms of their chemical constituents. It is generally assumed that the specific combination of secondary metabolites in the plant are responsible for the activity claimed or demonstrated, a concept called synergy.

Herbalists tend to use extracts from parts of plants, such as the roots or leaves but not isolate particular phytochemicals. Pharmaceutical medicine prefers single ingredients on the grounds that dosage can be more easily quantified. It is also possible to patent single compounds, and therefore generate income. Herbalists often reject the notion of a single active ingredient, arguing that the different phytochemicals present in many herbs will interact to enhance the therapeutic effects of the herb and dilute toxicity. Furthermore, they argue that a single ingredient may contribute to multiple effects. Herbalists deny that herbal synergism can be duplicated with synthetic chemicals They argue that phytochemical interactions and trace components may alter the drug response in ways that cannot currently be replicated with a combination of a few potentially active ingredients. Pharmaceutical researchers recognize the concept of drug synergism but note that clinical trials may be used to investigate the efficacy of a particular herbal preparation, provided the formulation of that herb is consistent.

In specific cases the claims of synergy and multifunctionality have been supported by science. The open question is how widely both can be generalized. Herbalists would argue that cases of synergy can be widely generalized, on the basis of their interpretation of evolutionary history, not necessarily shared by the pharmaceutical community. Plants are subject to similar selection pressures as humans and therefore they must develop resistance to threats such as radiation, reactive oxygen species and microbial attack in order to survive. Optimal chemical defenses have been selected for and have thus developed over millions of years. Human diseases are multifactorial and may be treated by consuming the chemical defences that they believe to be present in herbs. Bacteria, inflammation, nutrition and reactive oxygen species may all play a role in arterial disease. Herbalists claim a single herb may simultaneously address several of these factors. In short herbalists view their field as the study of a web of relationships rather than a quest for single cause and a single cure for a single condition.

In selecting herbal treatments herbalists may use forms of information that are not applicable to pharmacists. Because herbs can moonlight as vegetables, teas or spices they have a huge consumer base and large-scale epidemiological studies become feasible. Ethnobotanical studies are another source of information. Herbalists contend that historical medical records and herbals are underutilized resources. They favor the use of convergent information in assessing the medical value of plants. An example would be when in-vitro activity is consistent with traditional use.

Uses of Herbal Medicines by Animals

Indigenous healers often claim to have learned by observing that sick animals change their food preferences to nibble at bitter herbs they would normally reject. Field biologists have provided corroborating evidence based on observation of diverse species, such as chickens, sheep, butterflies, and chimpanzee. The habit has been shown to be a physical means of purging intestinal parasites. Lowland gorillas take 90% of their diet from the fruits of *Aframomum melegueta*, a relative of the ginger plant, that is a potent antimicrobial and apparently keeps shigellosis and similar infections at bay. Current research focuses on the possibility that this plants also protects gorillas from fibrosing cardiomyopathy which has a devastating effect on captive animals.

Sick animals tend to forage plants rich in secondary metabolites, such as tannins and alkaloids. Since these phytochemicals often have antiviral, antibacterial, antifungal and antihelminthic properties, a plausible case can be made for self-medication by animals in the wild.

History of Herbalism

The history of herbalism is closely tied with the history of medicine from prehistoric times up until the development of the germ theory of disease in the 19th century. Modern medicine from the 19th century to today has been based on evidence gathered using the scientific method. Evidence-based use of pharmaceutical drugs, often derived from medicinal plants, has largely replaced herbal treatments in modern health care. However, many people continue to employ various forms of traditional or alternative medicine. These systems often have a significant herbal component. The history of herbalism also overlaps with food history, as many of the herbs and spices historically used by humans to season food yield useful medicinal compounds, and use of spices with antimicrobial activity in cooking is part of an ancient response to the threat of food-borne pathogens.

Prehistory

The use of plants as medicines predates written human history. Archaeological evidence indicates that humans were using medicinal plants during the Paleolithic, approximately 60,000 years ago. (Furthermore, other non-human primates are also known to ingest medicinal plants to treat illness) Plant samples gathered from prehistoric burial sites have been thought to support the claim that Paleolithic peoples had knowledge of herbal medicine. For instance, a 60,000-year-old Neanderthal burial site, "Shanidar IV", in northern Iraq has yielded large amounts of pollen from 8 plant species, 7 of which are used now as herbal remedies. More recently Paul B. Pettitt has written that "A recent examination of the microfauna from the strata into which the grave was cut suggests that the pollen was deposited by the burrowing rodent Meriones tersicus, which is common in the Shanidar microfauna and whose burrowing activity can be observed today". Medicinal herbs were found in the personal effects of *Ötzi the Iceman*, whose body was frozen in the Ötztal Alps for more than 5,000 years. These herbs appear to have been used to treat the parasites found in his intestines.

Ancient History

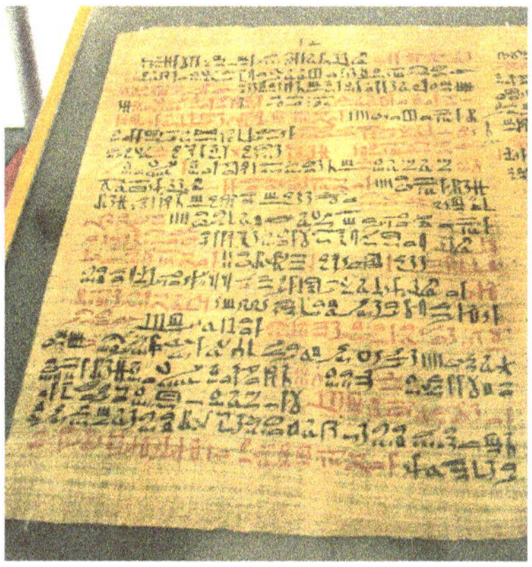

The Ebers Papyrus (c. 1550 BCE) from ancient Egypt has a prescription for *Cannabis sativa* (marijuana) applied topically for inflammation.

Mesopotamia

In Mesopotamia, the written study of herbs dates back over 5,000 years to the Sumerians, who created clay tablets with lists of hundreds of medicinal plants (such as myrrh and opium).

Ancient Egypt

Ancient Egyptian texts are of particular interest due to the language and translation controversies that accompany texts from this era and region. These differences in conclusions stem from the lack of complete knowledge of the Egyptian language: many translations are composed of mere approximations between Egyptian and modern ideas,and there can never be complete certainty of meaning or context. While physical documents are scarce, texts such as the Papyrus Ebers serve to illuminate and relieve some of the conjecture surrounding ancient herbal practices. The Papyrus consists of lists of ailments and their treatments, ranging from "disease of the limbs" to "diseases of the skin" and has information on over 850 plant medicines, including garlic, juniper, cannabis, castor bean, aloe, and mandrake. Treatments were mainly aimed at ridding the patient of the most prevalent symptoms because the symptoms were largely regarded as the disease itself. Knowledge of the collection and preparation of such remedies are mostly unknown, as many of the texts available for translation assume the physician already has some knowledge of how treatments are conducted and therefore such techniques would not need restating. Though modern understanding of Egyptian herbals stem form the translation of ancient texts, there is no doubt that trade and politics carried the Egyptian tradition to regions across the world, influencing and evolving many cultures medical practices and allowing for a glimpse into the world of ancient Egyptian medicine. Herbs used by Egyptian healers were mostly indigenous in origin, although some were imported from other regions like Lebanon. Other than papyri, evidence of herbal medicine has also been found in tomb illustrations or jars containing traces of herbs.

India

In India, Ayurveda medicine has used many herbs such as turmeric possibly as early as 4,000 BC. Earliest Sanskrit writings such as the Rig Veda, and Atharva Veda are some of the earliest available documents detailing the medical knowledge that formed the basis of the Ayurveda system. Many other herbs and minerals used in Ayurveda were later described by ancient Indian herbalists such as Charaka and Sushruta during the 1st millennium BC. The *Sushruta Samhita* attributed to Sushruta in the 6th century BC describes 700 medicinal plants, 64 preparations from mineral sources, and 57 preparations based on animal sources.

China

By 2,000 BC, Chinese started prescribing to herbal medicines. In China, seeds likely used for herbalism have also been found in the archaeological sites of Bronze Age China dating from the Shang Dynasty. The mythological Chinese emperor Shennong is said to have written the first Chinese pharmacopoeia, the "Shennong Ben Cao Jing". The "Shennong Ben Cao Jing" lists 365 medicinal plants and their uses - including Ephedra (the shrub that introduced the drug ephedrine to modern medicine), hemp, and chaulmoogra (one of the first effective treatments for leprosy). Succeeding generations augmented on the *Shennong Bencao Jing*, as in the *Yaoxing Lun* (*Treatise on the Nature of Medicinal Herbs*), a 7th-century Tang Dynasty treatise on herbal medicine.

Ancient Greece and Rome

Hippocrates

The *Hippocratic Corpus* serves as a collection of texts that are associated with the 'Father of Western Medicine', Hippocrates of Kos. Though the actual authorship of some of these texts is disputed, each reflects the general ideals put forth by Hippocrates and his followers. The recipes and remedies included in parts of the *Corpus* no doubt reveal popular and prevalent treatments of the early ancient Greek period.

Though any of the herbals included in the *Corpus* are similar to those practiced in the religious sectors of healing, they differ strikingly in the lack of rites, prayers, or chants used in the application of remedies. This distinction is truly indicative of the Hippocratic preference for logic and reason within the practices of medicine.

The ingredients mentioned in the *Corpus* consist of a myriad of herbs, both local to Greece and imported from exotic locales such as Arabia. While many imported goods would have been too expensive for common household use, some of the suggested ingredients include the more common and cheaper elderberries and St. John's Wort.

Galen

Galen of Pergamon, a Greek physician practicing in Rome, was certainly prolific in his attempt to write down his knowledge on all things medical – and in his pursuit, he wrote many texts regarding herbs and their properties, most notably his *Works of Therapeutics*. In this text, Galen outlines the merging of each discipline within medicine that combine to restore health and prevent disease.

While the subject of therapeutics encompasses a wide array of topics, Galen's extensive work in the humors and four basic qualities helped pharmacists to better calibrate their remedies for the individual person and their unique symptoms.

Diocles of Carystus

The writings of Diocles of Carystus were also extensive and prolific in nature. With enough prestige to be referred to as "the second Hippocrates", his advice in herbalism and treatment was to be taken seriously. Though the original texts no longer exist, many medical scholars throughout the ages have quoted Diocles rather extensively, and it is in these fragments that we gain knowledge of his writings. It is purported that Diocles actually wrote the first comprehensive herbal- this work then cited numerous times by contemporaries such as Galen, Celsus, and Soranus.

Pliny

In what is one of the first encyclopedic texts, Pliny the Elder's *Natural History* serves as a comprehensive guide to nature and also presents an extensive catalog of herbs valuable in medicine. With over 900 drugs and plants listed, Pliny's writings provide a very large knowledge base upon which we may learn more about ancient herbalism and medical practices. Pliny himself referred to ailments as "the greatest of all the operations of nature," and the act of treatment via drugs as impacting the "state of peace or of war which exists between the various departments of nature".

Dioscordes

Much like Pliny, Pedanius Dioscorides constructed a pharmacopeia, *De Materia Medica*, consisting of over 1000 medicines produced form herbs, minerals, and animals. The remedies that comprise this work were widely utilized throughout the ancient period and Dioscorides remained the greatest expert on drugs for over 1,600 years.

Similarly important for herbalists and botanists of later centuries was Theophrastus' *Historia Plantarum*, written in the 4th century BC, which was the first systematization of the botanical world.

Middle Ages

While there are certainly texts from the medieval period that denote the uses of herbs, there has been a long-standing debate between scholars as to the actual motivations and understandings that underline the creation of herbal documents during the medieval period. The first point of view dictates that the information presented in these medieval texts were merely copied from their classical equivalents without much thought or understanding. The second viewpoint, which is gaining traction among modern scholars, states that herbals were copied for actual use and backed by genuine understanding.

Some evidence for the suggestion that herbals were utilized with knowledgeable intent, was the addition of several chapters of plants, lists of symptoms, habitat information, and plant synonyms added to texts such as the *Herbarium*. Notable texts utilized in this time period include Bald's

Leechbook, the Lacnunga, the peri didaxeon, Herbarium Apulei, Da Taxone, and Madicina de Quadrupedidus, while the most popular during this time period were the Ex Herbis Femininis, the Herbarius, and works by Dioscorides.

Benedictine monasteries were the primary source of medical knowledge in Europe and England during the Early Middle Ages. However, most of these monastic scholars' efforts were focused on translating and copying ancient Greco-Roman and Arabic works, rather than creating substantial new information and practices. Many Greek and Roman writings on medicine, as on other subjects, were preserved by hand copying of manuscripts in monasteries. The monasteries thus tended to become local centers of medical knowledge, and their herb gardens provided the raw materials for simple treatment of common disorders. At the same time, folk medicine in the home and village continued uninterrupted, supporting numerous wandering and settled herbalists. Among these were the "wise-women" and "wise men", who prescribed herbal remedies often along with spells, enchantments, divination and advice. It was not until the late Middle Ages that women and men who were knowledgeable in herb lore became the targets of the witch hysteria. One of the most famous women in the herbal tradition was Hildegard of Bingen. A 12th-century Benedictine nun, she wrote a medical text called *Causae et Curae*. During this time, herbalism was mainly practiced by women, particularly among Germanic tribes.

There were three major sources of information on healing at the time including the Arabian School, Anglo-Saxon leechcraft, and Salerno. A great scholar of the Arabian School was Avicenna, who wrote *The Canon of Medicine* which became the standard medical reference work of the Arab world. "*The Canon of Medicine* is known for its introduction of systematic experimentation and the study of physiology, the discovery of contagious diseases and sexually transmitted diseases, the introduction of quarantine to limit the spread of infectious diseases, the introduction of experimental medicine, clinical trials, and the idea of a syndrome in the diagnosis of specific diseases. ...*The Canon* includes a description of some 760 medicinal plants and the medicine that could be derived from them." With Leechcraft, though bringing to mind part of their treatments, leech was the English term for medical practitioner. Salerno was a famous school in Italy centered around health and medicine. A student of the school was Constantine the African, credited with bringing Arab medicine to Europe.

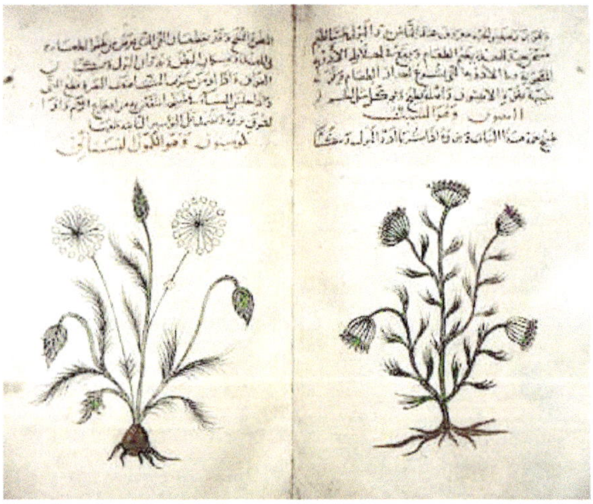

Dioscorides' *Materia Medica*, c. 1334 copy in Arabic, describes medicinal features of cumin and dill.

Translation of Herbals

During the Middle Ages, the study of plants began to be based on critical observations. "In the 16th and 17th century an interest in botany revived in Europe and spread to America by way of European conquest and colonization." Philosophers started to act as herbalists and academic professors studied plants with great depth. Herbalists began to explore the use of plants for both medicinal purposes and agricultural uses. Botanists in the Middle Ages were known as herbalists; they collected, grew, dried, stored, and sketched plants. Many became experts in identifying and describing plants according to their morphology and habitats, as well as their usefulness. These books, called herbals included beautiful drawings and paintings of plants as well as their uses.

At that time both botany and the art of gardening stressed the utility of plants for man; the popular herbal, described the medical uses of plants. During the Middle Ages, there was an expansion of book culture that spread through the medieval world. The phenomenon of translation is well-documented, from its beginnings as a scholarly endeavor in Baghdad as early as the eighth century to its expansion throughout European Mediterranean centers of scholarship by the eleventh and twelfth centuries. The process of translation is collaborative effort, requiring a variety of people to translate and add to them. However, how the Middle Ages viewed nature seems to be a mystery.

Translation of text and image has provided numerous versions and compilations of individual manuscripts from diverse sources, old and new. Translation is a dynamic process as well as a scholarly endeavor that contributed great to science in the Middle Ages; the process naturally entailed continuous revisions and additions. The Benedictine monasteries were known for their in-depth knowledge of herbals. These gardens grew the herbs which were considered to be useful for the treatment of the various human ills; the beginnings of modern medical education can be connected with monastic influence. Monastic academies were developed and monks were taught how to translate Greek manuscripts into Latin.

Knowledge of medieval botanicals was closely related to medicine because the plant's principal use was for remedies. Herbals were structured by the names of the plants, identifying features, medicinal parts of plant, therapeutic properties, and some included instructions on how to prepare and use them. For medical use of herbals to be effective, a manual was developed. Dioscorides' *De material medica* was a significant herbal designed for practical purposes.

Theophrastus wrote more than 200 papers describing the characteristics of over 500 plants. He developed a classification system for plants based on their morphology such as their form and structure. He described in detail pepper, cinnamon, bananas, asparagus, and cotton. Two of his best-known works, *Enquiry into Plants* and *The Causes of Plants*, have survived for many centuries and were translated into Latin. He has been referred to as the "grandfather of botany". Crateuas was the first to produce a pharmacological book for medicinal plants, and his book influenced medicine for many centuries. A Greek physician, Pedanius Dioscorides described over 600 different kinds of plants and describes their useful qualities for herbal medicine, and his illustrations were used for pharmacology and medicine as late as the Renaissance years.

Monasteries established themselves as centers for medical care. Information on these herbals and how to use them was passed on from monks to monks, as well as their patients. These illustrations were of no use to everyday individuals; they were intended to be complex and for people with prior knowledge and understanding of herbal. The usefulness of these herbals have been questioned because they appear to be unrealistic and several plants are depicted claiming to cure the same condition, as "the modern world does not like such impression." When used by experienced healers, these plants can provide their many uses. For these medieval healers, no direction was needed their background allowed them to choose proper plants to use for a variety of medical conditions. The monk's purpose was to collect and organize text to make them useful in their monasteries. Medieval monks took many remedies from classical works and adapted them to their own needs as well as local needs. This may be why none of the collections of remedies we have presently agrees fully with another.

Another form of translation was oral transmission; this was used to pass medical knowledge from generation to generation. A common misconception is that one can know early medieval medicine simply by identifying texts, but it is difficult to compose a clear understanding of herbals without prior knowledge. There are many factors that played in influenced in the translation of these herbals, the act of writing or illustrating was just a small piece of the puzzle, these remedies stems from many previous translations the incorporated knowledge from a variety of influences.

Early Modern Era

The 16th and 17th centuries were the great age of herbals, many of them available for the first time in English and other languages rather than Latin or Greek. The 18th and 19th centuries saw more incorporation of plants found in the Americas, but also the advance of modern medicine.

16th Century

The first herbal to be published in English was the anonymous *Grete Herball* of 1526. The two best-known herbals in English were *The Herball or General History of Plants* (1597) by John Gerard and *The English Physician Enlarged* (1653) by Nicholas Culpeper. Gerard's text was basically a pirated translation of a book by the Belgian herbalist Dodoens and his illustrations came from a German botanical work. The original edition contained many errors due to faulty matching of the two parts. Culpepper's blend of traditional medicine with astrology, magic, and folklore was ridiculed by the physicians of his day, yet his book - like Gerard's and other herbals - enjoyed phenomenal popularity. The Age of Exploration and the Columbian Exchange introduced new medicinal plants to Europe. The *Badianus Manuscript* was an illustrated Mexican herbal written in Nahuatl and Latin in the 16th century.

17th Century

The second millennium, however, also saw the beginning of a slow erosion of the pre-eminent position held by plants as sources of therapeutic effects. This began with the Black Death, which the then dominant Four Element medical system proved powerless to stop. A century later, Paracelsus introduced the use of active chemical drugs (like arsenic, copper sulfate, iron, mercury, and sulfur).

18th Century

In the Americas, herbals were relied upon for most medical knowledge with physicians being few and far between. These books included almanacs, Dodoens' *New Herbal, Edinburgh New Dis-*

pensatory, Buchan's *Domestic Medicine,* and other works. Aside from European knowledge on American plants, Native Americans shared some of their knowledge with colonists, but most of these records were not written and compiled until the 19th century. John Bartram was a botanist that studied the remedies that Native Americans would share and often included bits of knowledge of these plants in printed almanacs.

19th Century

The formalization of pharmacology in the 19th century led to greater understanding of the specific actions drugs have on the body. At that time, Samuel Thompson was an uneducated but well respected herbalist who influenced professional opinions so much that Doctors and Herbalists would refer to themselves as Thompsonians. They distinguished themselves from "regular" doctors of the time who used calomel and bloodletting, and led to a brief renewal of the empirical method in herbal medicine.

Modern Era

Traditional herbalism has been officially regarded as a method of alternative medicine in the United States since the Flexner Report of 1910 led to the closing of the eclectic medical schools where botanical medicine was exclusively practiced. In China, Mao Zedong reintroduced Traditional Chinese Medicine, which relied heavily on herbalism, into the health care system in 1949. Since then, schools have been training thousands of practitioners – including Americans – in the basics of Chinese medicines to be used in hospitals. "The Chinese pharmacopeia is 5000 years old and contains substances that are transformed into effective but non-toxic medicines." While Britain in the 1930s was experiencing turbulence over the practice of herbalism, in the United States, government regulation began to prohibit the practice.

"The World Health Organization estimated that 80% of people worldwide rely on herbal medicines for some part of their primary health care. In Germany, about 600 to 700 plant based medicines are available and are prescribed by some 70% of German physicians." South African doctors use herbs to treat many patients, some patients finding the plants more useful than modern medicine previously prescribed to them.

The practice of prescribing treatments and cures to patients requires a legal medical license in the United States of America, and the licensing of these professions occurs on a state level. "There is currently no licensing or certification for herbalists in any state that precludes the rights of anyone to use, dispense, or recommend herbs."

"Traditional medicine is a complex network of interaction of both ideas and practices, the study of which requires a multidisciplinary approach." Many alternative physicians in the 21st century incorporate herbalism in traditional medicine due to the diverse abilities plants have and their low number of side effects.

Herbal Distillate

Herbal distillates, also known as floral waters, hydrosols, hydrolates, herbal waters, and essential waters, are aqueous products of hydro distillation. They are colloidal suspensions of essential oils as well as water-soluble components obtained by steam distillation or hydrodistillation (a variant of steam distillation) from plants/herbs. These herbal distillates have uses as flavorings and cosmetics (skin care).

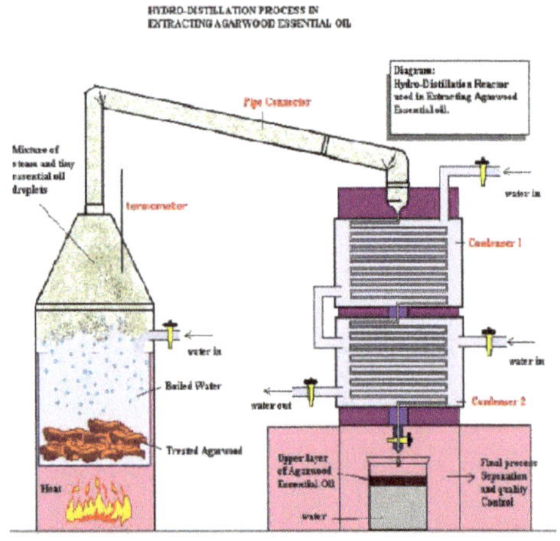

Hydro-Distillation Process in Extracting Of Agarwood Essential Oil.

Production

Herbal distillates are produced in the same or similar manner as essential oils. However, the essential oil will float to the top of the distillate where it is removed, leaving behind the watery distillate. For this reason perhaps the term *essential water* is more descript. In the past, these essential waters were often considered a byproduct of distillation, but now are considered an important co-product. The produced herbal waters are essentially diluted essential oils at less than 1% concentration (typically 0.01% to 0.04%).

Usage

Cosmetics and toiletries makers are finding many uses for herbal distillates. Distillates are also used as flavorings and curables.

Science

The science of distillation is based on the fact that different substances vaporise at different temperatures. Unlike other extraction techniques based on solubility of a compound in either water or oil, distillation will separate components regardless of their solubility. The distillate will contain compounds that vaporize at or below the temperature of distillation. The actual chemical components of these orange herbal distillates have not yet been fully identified, but plant distillates will usually contain essential oil compounds as well as organic acids and other water-soluble plant components. Compounds with a higher vaporization point will remain behind and will include many of the water-soluble plant pigments and flavonoids.

Because hydrosols are produced at high temperatures and are somewhat acidic, they tend to inhibit bacterial growth but not fungal growth. They are not sterile. They are a fresh product, like milk, and should be kept refrigerated. Small-scale producers of hydrosols must be particularly aware of, and take steps to prevent bacterial contamination.

Cannabis (Drug)

Cannabis also known as marijuana among several other names, is a preparation of the *Cannabis* plant intended for use as a psychoactive drug or medicine. The main psychoactive part of cannabis is tetrahydrocannabinol (THC); one of 483 known compounds in the plant, including at least 65 other cannabinoids. Cannabis can be used by smoking, vaporization, within food, or as an extract.

Cannabis is often used for its mental and physical effects, such as a "high" or "stoned" feeling, a general change in perception, euphoria (heightened mood), and an increase in appetite. Short term side effects may include a decrease in short-term memory, dry mouth, impaired motor skills, red eyes, and feelings of paranoia or anxiety. Long term side effects may include addiction, decreased mental ability in those who started as teenagers, and behavioral problems in children whose mothers used cannabis during pregnancy. Onset of effects is within minutes when smoked and about 30 to 60 minutes when cooked and eaten. They last for between two and six hours.

Cannabis is mostly used recreationally or as a medicinal drug. It may also be used for religious or spiritual purposes. In 2013, between 128 and 232 million people used cannabis (2.7% to 4.9% of the global population between the ages of 15 and 65). In 2015, 43% of Americans had used cannabis, which increased to 51% in 2016. About 12% have used it in the past year, and 7.3% have used it in the past month. This makes it the most commonly used illegal drug both in the world and the United States.

The earliest recorded uses date from the 3rd millennium BC. Since the early 20th century, cannabis has been subject to legal restrictions. The possession, use, and sale of cannabis is illegal in most countries of the world. Medical cannabis refers to the physician-recommended use of cannabis, which is taking place in Canada, Belgium, Australia, the Netherlands, Spain, and 23 U.S. states. Cannabis use started to become popular in the US in the 1970s. Support for legalization has increased in the United States and several US states have legalized recreational or medical use.

Uses

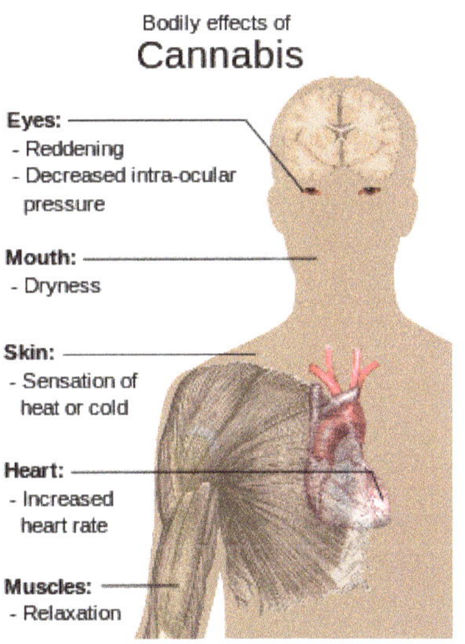

Main short-term physical effects of cannabis

Medical

Medical cannabis, or medical marijuana can refer to the use of cannabis and its cannabinoids to treat disease or improve symptoms; however, there is no single agreed upon definition. The use of cannabis as a medicine has not been rigorously scientifically tested, often due to production restrictions and other federal regulations. There is limited evidence suggesting cannabis can be used to reduce nausea and vomiting during chemotherapy, to improve appetite in people with HIV/AIDS, and to treat chronic pain and muscle spasms. Its use for other medical applications is insufficient for conclusions about safety or efficacy.

Short-term use increases the risk of both minor and major adverse effects. Common side effects include dizziness, feeling tired, vomiting, and hallucinations. Long-term effects of cannabis are not clear. Concerns include memory and cognition problems, risk of addiction, schizophrenia in young people, and the risk of children taking it by accident.

Recreational

A woman smoking a marijuana "joint".

Cannabis has psychoactive and physiological effects when consumed. The immediate desired effects from consuming cannabis include relaxation and euphoria (the "high" or "stoned" feeling), a general alteration of conscious perception, increased awareness of sensation, increased libido and distortions in the perception of time and space. At higher doses, effects can include altered body image, auditory and/or visual illusions, pseudohallucinations and ataxia from selective impairment of polysynaptic reflexes. In some cases, cannabis can lead to dissociative states such as depersonalization and derealization.

Some immediate undesired side effects include a decrease in short-term memory, dry mouth, impaired motor skills and reddening of the eyes. Aside from a subjective change in perception and mood, the most common short-term physical and neurological effects include increased heart rate, increased appetite and consumption of food, lowered blood pressure, impairment of short-term and working memory, psychomotor coordination, and concentration. Some users may experience an episode of acute psychosis, which usually abates after six hours, but in rare instances, heavy users may find the symptoms continuing for many days. A reduced quality of life is associated with heavy cannabis use, although the relationship is inconsistent and weaker than for tobacco and other substances.

Spiritual

Process of making bhang in a Sikh village in Punjab, India. On the festival of colors called Holi, it is a customary addition to some intoxicating drinks.

Cannabis has held sacred status in several religions. It has been used in an entheogenic context – a chemical substance used in a religious, shamanic, or spiritual context - in India and Nepal since the Vedic period dating back to approximately 1500 BCE, but perhaps as far back as 2000 BCE. There are several references in Greek mythology to a powerful drug that eliminated anguish and sorrow. Herodotus wrote about early ceremonial practices by the Scythians, thought to have occurred from the 5th to 2nd century BCE. In modern culture the spiritual use of cannabis has been spread by the disciples of the Rastafari movement who use cannabis as a sacrament and as an aid to meditation. The earliest known reports regarding the sacred status of cannabis in India and Nepal come from the Atharva Veda estimated to have been written sometime around 2000–1400 BCE.

Available Forms

A joint prior to rolling, with a paper handmade filter on the left

Cannabis is consumed in many different ways:

- smoking, which typically involves burning and inhaling vaporized cannabinoids ("smoke") from small pipes, bongs (portable versions of hookahs with a water chamber), paper-wrapped joints or tobacco-leaf-wrapped blunts, and other items.
- vaporizer, which heats any form of cannabis to 165–190 °C (329–374 °F), causing the active ingredients to evaporate into a vapor without burning the plant material (the boiling point of THC is 157 °C (315 °F) at 760 mmHg pressure).

- cannabis tea, which contains relatively small concentrations of THC because THC is an oil (lipophilic) and is only slightly water-soluble (with a solubility of 2.8 mg per liter). Cannabis tea is made by first adding a saturated fat to hot water (e.g. cream or any milk except skim) with a small amount of cannabis.
- edibles, where cannabis is added as an ingredient to one of a variety of foods, including butter and baked goods. In India it is commonly made into a beverage, bhang.

Adverse Effects

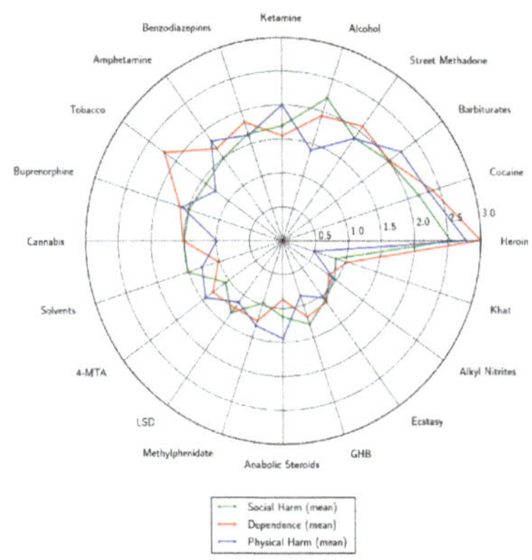

Addiction experts in psychiatry, chemistry, pharmacology, forensic science, epidemiology, and the police and legal services engaged in delphic analysis regarding 20 popular recreational drugs. Cannabis was ranked 11th in dependence, 17th in physical harm, and 10th in social harm.

According to the United States Department of Health and Human Services, there were 455,000 emergency room visits associated with cannabis use in 2011. These statistics include visits in which the patient was treated for a condition induced by or related to recent cannabis use. The drug use must be "implicated" in the emergency department visit, but does not need to be the direct cause of the visit. Most of the illicit drug emergency room visits involved multiple drugs. In 129,000 cases, cannabis was the only implicated drug.

Heavy, long term exposure to marijuana may have biologically-based physical, mental, behavioral and social health consequences and may be "associated with diseases of the liver (particularly with co-existing hepatitis C), lungs, heart, and vasculature". It is recommended that cannabis use be stopped before and during pregnancy as it can result in negative outcomes for both the mother and baby. However, maternal use of marijuana during pregnancy does not appear to be associated with low birth weight or early delivery after controlling for tobacco use and other confounding factors. A 2014 review found that while cannabis use may be less harmful than alcohol use, the recommendation to substitute it for problematic drinking is premature without further study.

Toxicity

THC, the principal psychoactive constituent of the cannabis plant, has low toxicity. The dose of THC needed to kill 50% of tested rodents is extremely high. Acute effects may include anxiety and panic, impaired attention, and memory (while intoxicated), an increased risk of psychotic symptoms, and possibly an increased risk of accidents if a person drives a motor vehicle while intoxicated. Short-term cannabis intoxication can hinder the mental processes of organizing and collecting thoughts. This condition is known as temporal disintegration. Psychotic episodes are well-documented and typically resolve within minutes or hours. There have been few reports of symptoms lasting longer. Cannabis has not been reported to cause fatal overdose. Studies have found that cannabis use during adolescence is associated with impairments in memory that persist beyond short-term intoxication.

Lungs

A limited number of studies have examined the effects of cannabis smoking on the respiratory system. Chronic heavy marijuana smoking is associated with coughing, production of sputum, wheezing, and other symptoms of chronic bronchitis. The available evidence does not support a causal relationship between cannabis use and chronic obstructive pulmonary disease. Short-term use of cannabis is associated with bronchodilation.

Cancer

Cannabis smoke contains thousands of organic and inorganic chemical compounds. This tar is chemically similar to that found in tobacco smoke, and over fifty known carcinogens have been identified in cannabis smoke, including; nitrosamines, reactive aldehydes, and polycylic hydrocarbons, including benz[a]pyrene. Cannabis smoke is also inhaled more deeply than is tobacco smoke. As of 2015, there is no consensus regarding whether cannabis smoking is associated with an increased risk of cancer. Light and moderate use of cannabis is not believed to increase risk of lung or upper airway cancer. Evidence for causing these cancers is mixed concerning heavy, long-term use. In general there are far lower risks of pulmonary complications for regular cannabis smokers when compared with those of tobacco. A 2015 review found an association between cannabis use and the development of testicular germ cell tumors (TGCTs), particularly non-seminoma TGCTs. A 2015 analysis of six studies found little evidence that long-term or regular cannabis smoking was associated with lung cancer risk, though it could not rule out whether an association with heavy smoking exists. Another 2015 meta-analysis found no association between lifetime cannabis use and risk of head or neck cancer. Combustion products are not present when using a vaporizer, consuming THC in pill form, or consuming cannabis foods.

Cardiovascular

There is serious suspicion among cardiologists, spurring research but falling short of definitive proof, that cannabis use has the potential to contribute to cardiovascular disease. Cannabis is believed to be an aggravating factor in rare cases of arteritis, a serious condition that in some cases leads to amputation. Because 97% of case-reports also smoked tobacco, a formal association with cannabis could not be made. If cannabis arteritis turns out to be a distinct clinical entity, it might be the consequence of vasoconstrictor activity observed from delta-8-THC and delta-9-THC. Other serious cardiovascular events including myocardial infarction, stroke, sudden cardiac death, and cardiomyopathy have been reported to be temporally associated with cannabis use. Research in these events is complicated because cannabis is often used in conjunction with tobacco, and drugs such as alcohol and cocaine. These putative effects can be taken in context of a wide range of cardiovascular phenomena regulated by the endocannabinoid system and an overall role of cannabis in causing decreased peripheral resistance and increased cardiac output, which potentially could pose a threat to those with cardiovascular disease. There is some evidence from case reports that cannabis use may provoke fatal cardiovascular events in young people who have not been diagnosed with cardiovascular disease. Smoking cannabis has also been shown to increase the risk of myocardial infarction by 4.8 times for the 60 minutes after consumption.

Neurological

Anatomy and Brain Chemistry

Cannabis use is associated with neuroanatomic alterations in brain regions rich in cannabinoid receptors, such as the hippocampus, prefrontal cortex, amygdala, and cerebellum. The same review found that greater dose of marijuana and earlier age at onset of use were also associated with such alterations. A 2010 review found resting blood flow to be lower globally and in prefrontal areas of the brain in cannabis users, when compared to non-users. It was also shown that giving THC or cannabis correlated with increased bloodflow in these areas, and facilitated activation of the anterior cingulate cortex and frontal cortex when participants were presented with assignments demanding use of cognitive capacity. Both reviews noted that some of the studies that they examined had methodological limitations, for example small sample sizes or not distinguishing adequately between cannabis and alcohol consumption. Cannabis users appear to have smaller hippocampi than nonusers; this finding is based on a series of small studies with inconsistent designs, so it is uncertain. A 2016 meta-analysis found that regular cannabis users tended to have cue reactivity, the intensity of which ranged from moderate to very high.

There is limited evidence that chronic cannabis use can reduce levels of glutamate metabolites in the human brain.

Function

A 2012 meta-analysis found that the effects of cannabis use on neurocognitive functions were "limited to the first 25 days of abstinence" and that there was no evidence that such use had long-lasting effects, while a 2011 review found that cannabis use impaired cognitive functions on several levels, ranging from basic coordination to executive function tasks.

Psychiatric

Epidemiological studies have found a strong correlation between cannabis use and the risk of psychosis, and found the risk is higher for with strains higher in THC. However, not all researchers consider this association to reflect a cause-and-effect relationship between cannabis use and psychosis.

It is not clear whether cannabis use affects the rate of suicide. It may increase the risk of depression, but further research is needed in this area.

Chronic Use

Effects of chronic use may include bronchitis, a cannabis dependence syndrome, and subtle impairments of attention and memory. These deficits persist while chronically intoxicated. There is little evidence that cognitive impairments persist in adult abstinent cannabis users. Compared to non-smokers, people who smoked cannabis regularly in adolescence exhibit reduced connectivity in specific brain regions associated with memory, learning, alertness, and executive function. A study has suggested that sustained heavy, daily, adolescent onset cannabis use over decades is associated with a decline in IQ by age 38. No effects were found in those who initiated cannabis use later, or in those who ceased use earlier in adulthood.

Tolerance and withdrawal

Cannabis usually causes no tolerance or withdrawal symptoms except in heavy users. In a survey of heavy users, 42.4% experienced withdrawal symptoms when they tried to quit marijuana, such as craving, irritability, boredom, anxiety and sleep disturbances. About 9% of those who experiment with marijuana eventually become dependent. The rate goes up to 1 in 6 among those who begin use as adolescents, and one-quarter to one-half of those who use it daily according to a NIDA review. A 2013 review estimates daily use is associated with a 10-20% rate of dependence. The highest risk of cannabis dependence is found in those with a history of poor academic achievement, deviant behavior in childhood and adolescence, rebelliousness, poor parental relationships, or a parental history of drug and alcohol problems. Cannabis withdrawal is less severe than withdrawal from alcohol.

Motor vehicle crashes

Marijuana is the most common illegal drug reported in motor vehicle accidents. A 2012 meta-analysis found that cannabis use was associated with an increased risk of being involved in a motor vehicle crash. A 2016 review also found a statistically significant increase in crash risk associated with marijuana use, but noted that this risk was "of low to medium magnitude." The increase in risk of motor vehicle crash for cannabis use is between 2 and 3 times relative to baseline, whereas that for comparable doses of alcohol is between 6 and 15 times.

Pharmacology

Mechanism of Action

The high lipid-solubility of cannabinoids results in their persisting in the body for long periods of time. Even after a single administration of THC, detectable levels of THC can be found in the body for weeks or longer (depending on the amount administered and the sensitivity of the assessment method). A number of investigators have suggested that this is an important factor in marijuana's effects, perhaps because cannabinoids may accumulate in the body, particularly in the lipid membranes of neurons.

Not until the end of the 20th century was the specific mechanism of action of THC at the neuronal level studied. Researchers have subsequently confirmed that THC exerts its most prominent effects via its actions on two types of cannabinoid receptors, the CB_1 receptor and the CB_2 receptor, both of which are G-protein coupled receptors. The CB1 receptor is found primarily in the brain as well as in some peripheral tissues, and the CB2 receptor is found primarily in peripheral tissues, but is also expressed in neuroglial cells. THC appears to alter mood and cognition through its agonist actions on the CB1 receptors, which inhibit a secondary messenger system (adenylate cyclase) in a dose dependent manner. These actions can be blocked by the selective CB1 receptor antagonist SR141716A (rimonabant), which has been shown in clinical trials to be an effective treatment for smoking cessation, weight loss, and as a means of controlling or reducing metabolic syndrome risk factors. However, due to the dysphoric effect of CB1 antagonists, this drug is often discontinued due to these side effects.

Via CB1 activation, THC indirectly increases dopamine release and produces psychotropic effects. Cannabidiol also acts as an allosteric modulator of the mu and delta opioid receptors. THC also potentiates the effects of the glycine receptors. The role of these interactions in the "marijuana high" remains elusive.

Physical and Chemical Properties

Detection in Body Fluids

THC and its major (inactive) metabolite, THC-COOH, can be measured in blood, urine, hair, oral fluid or sweat using chromatographic techniques as part of a drug use testing program or a forensic investigation of a traffic or other criminal offense. The concentrations obtained from such analyses can often be helpful in distinguishing active use from passive exposure, elapsed time since use, and extent or duration of use. These tests cannot, however, distinguish authorized cannabis smoking for medical purposes from unauthorized recreational smoking. Commercial cannabinoid immunoassays, often employed as the initial screening method when testing physiological specimens for marijuana presence, have different degrees of cross-reactivity with THC and its metabolites. Urine contains predominantly THC-COOH, while hair, oral fluid and sweat contain primarily THC. Blood may contain both substances, with the relative amounts dependent on the recency and extent of usage.

The Duquenois–Levine test is commonly used as a screening test in the field, but it cannot definitively confirm the presence of cannabis, as a large range of substances have been shown to give false positives. Despite this, it is common in the United States for prosecutors to seek plea bargains on the basis of positive D–L tests, claiming them to be conclusive, or even to seek conviction without the use of gas chromatography confirmation, which can only be done in the lab. In 2011, researchers at John Jay College of Criminal Justice reported that dietary zinc supplements can mask the presence of THC and other drugs in urine. However, a 2013 study conducted by researchers at the University of Utah School of Medicine refute the possibility of self-administered zinc producing false-negative urine drug tests.

Varieties and Strains

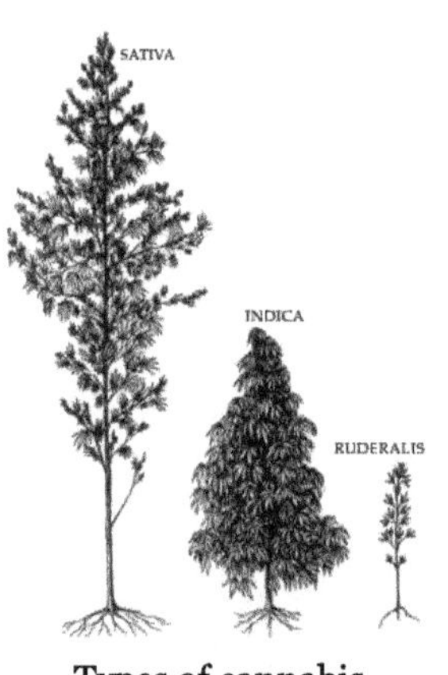

Types of cannabis

CBD is a 5-HT_{1A} receptor agonist, which may also contribute to an anxiolytic effect. This likely means the high concentrations of CBD found in *Cannabis indica* mitigate the anxiogenic effect of THC significantly. The effects of sativa are well known for their cerebral high, hence its daytime use as medical cannabis, while indica is well known for its sedative effects and preferred night time use as medical cannabis.

Psychoactive Ingredients

According to the United Nations Office on Drugs and Crime (UNODC), "the amount of THC present in a cannabis sample is generally used as a measure of cannabis potency." The three main forms of cannabis products are the flower, resin (hashish), and oil (hash oil). The UNODC states that cannabis often contains 5% THC content, resin "can contain up to 20% THC content", and that "Cannabis oil may contain more than 60% THC content."

A 2012 review found that the THC content in marijuana had increased worldwide from 1970 to 2009. It is unclear, however, whether the increase in THC content has caused people to consume more THC or if users adjust based on the potency of the cannabis. It is likely that the higher THC content allows people to ingest less tar. At the same time, Cannabidiol (CBD) levels in seized samples have lowered, in part because of the desire to produce higher THC levels and because more illegal growers cultivate indoors using artificial lights. This helps avoid detection but reduces the CBD production of the plant.

Australia's National Cannabis Prevention and Information Centre (NCPIC) states that the buds (flowers) of the female cannabis plant contain the highest concentration of THC, followed by the leaves. The stalks and seeds have "much lower THC levels". The UN states that leaves can contain ten times less THC than the buds, and the stalks one hundred times less THC.

After revisions to cannabis rescheduling in the UK, the government moved cannabis back from a class C to a class B drug. A purported reason was the appearance of high potency cannabis. They believe skunk accounts for between 70 and 80% of samples seized by police (despite the fact that skunk can sometimes be incorrectly mistaken for all types of herbal cannabis). Extracts such as hashish and hash oil typically contain more THC than high potency cannabis flowers.

Preparations

Dried flower buds

Kief

Hashish

Pipe resin

Marijuana

Marijuana or marihuana (herbal cannabis), consists of the dried flowers and subtending leaves and stems of the female *Cannabis* plant. This is the most widely consumed form, containing 3% to 20% THC, with reports of up-to 33% THC. This is the stock material from which all other preparations are derived. Although herbal cannabis and industrial hemp derive from the same species and contain the psychoactive component (THC), they are distinct strains with unique biochemical compositions and uses. Hemp has lower concentrations of THC and higher concentrations of cannabidiol (CBD), which decreases the psychoactive effects

Kief

Kief is a powder, rich in trichomes, which can be sifted from the leaves and flowers of cannabis plants and either consumed in powder form or compressed to produce cakes of hashish. The word "kif" derives from colloquial Arabic meaning *pleasure*.

Hashish

Hashish (also spelled hasheesh, hashisha, or simply hash) is a concentrated resin cake or ball produced from pressed kief, the detached trichomes and fine material that falls off cannabis flowers and leaves. or from scraping the resin from the surface of the plants and rolling it into balls. It varies in color from black to golden brown depending upon purity and variety of cultivar it was obtained from. It can be consumed orally or smoked, and is also vaporised, or 'vaped'. The term "Rosin Hash" refers to a high quality solventless product obtained through heat and pressure.

Tincture

Cannabinoids can be extracted from cannabis plant matter using high-proof spirits (often grain alcohol) to create a tincture, often referred to as "green dragon". Nabiximols is a branded product name from a tincture manufacturing pharmaceutical company.

Hash Oil

Hash oil is a resinous matrix of cannabinoids obtained from the *Cannabis* plant by solvent extraction, formed into a hardened or viscous mass. Hash oil can be the most potent of the main cannabis products because of its high level of psychoactive compound per its volume, which can vary depending on the plant's mix of essential oils and psychoactive compounds. Butane and supercritical carbon dioxide hash oil have become popular in recent years.

Infusions

There are many varieties of cannabis infusions owing to the variety of non-volatile solvents used. The plant material is mixed with the solvent and then pressed and filtered to express the oils of the plant into the solvent. Examples of solvents used in this process are cocoa butter, dairy butter, cooking oil, glycerine, and skin moisturizers. Depending on the solvent, these may be used in cannabis foods or applied topically.

Adulterated Cannabis

Contaminants or adulterants may be found in marijuana or hashish. Other substances may be added to cannabis to add weight to the product (lead has been used in some cases), to increase its psychoactive effects (e.g., Phencyclidine), or as part of the cultivation and processing of the cannabis (e.g., fertilizer). Hashish obtained from "soap bar"-type sources. The dried flowers of the plant may be contaminated by the plant taking up heavy metals and other toxins from its growing environment, or by the addition of glass. In the Netherlands, chalk has been used to make cannabis appear to be of a higher quality. Increasing the weight of hashish products in Germany with lead caused lead intoxication in at least 29 users.

Despite cannabis being generally perceived as a natural product, in a recent Australian survey one in four Australians consider cannabis grown indoors under hydroponic conditions to be a greater health risk due to increased contamination, added to the plant during cultivation to enhance the plant growth and quality.

Drug dealers may "spike" or lace marijuana with other chemicals such as PCP, creating a product known as "wet marijuana"; this enhances the effects of smoking it and it can be used to make low-grade, low-potency marijuana seem more effective.

Medical Use

Medical marijuana refers to the use of the *Cannabis* plant as a physician-recommended herbal therapy as well as synthetic THC and cannabinoids. So far, the medical use of cannabis is legal only in a limited number of territories, including Canada, Belgium, Australia, the Netherlands, Spain, and several U.S. states. This usage generally requires a prescription, and distribution is usually done within a framework defined by local laws. There is evidence supporting the use of cannabis or its derivatives in the treatment of chemotherapy-induced nausea and vomiting, neuropathic pain, and multiple sclerosis. Lower levels of evidence support its use for AIDS wasting syndrome, epilepsy, rheumatoid arthritis, and glaucoma.

History

麻

The Chinese character for hemp (*má*) depicts two plants under a shelter.
Cannabis cultivation dates back at least 3000 years in Taiwan.

Cannabis is indigenous to Central and South Asia. There is evidence of inhalation of cannabis smoke from the 3rd millennium BCE, namely charred cannabis seeds found in a ritual brazier at an ancient burial site in present-day Romania. The earliest written reference to cannabis dates back to 2727 B.C., from the Chinese emperor Shennong.

In 2003, a leather basket filled with cannabis leaf fragments and seeds was found next to a 2,500- to 2,800-year-old mummified shaman in the northwestern Xinjiang Uygur Autonomous Region of China. Evidence of cannabis consumption was also found in Egyptian mummies dated about 950 BC.

Hemp is called *ganja* (Sanskrit: IAST: gañjā) in Sanskrit and other modern Indo-Aryan languages. Some scholars suggest that the ancient drug soma, mentioned in the Vedas, was cannabis, although this theory is disputed.

Cannabis was also known to the ancient Assyrians, who discovered its psychoactive properties through the Aryans. Using it in some religious ceremonies, they called it *qunubu* (meaning "way to produce smoke"), a probable origin of the modern word "cannabis". The Aryans also introduced cannabis to the Scythians, Thracians and Dacians, whose shamans (the *kapnobatai*—"those who walk on smoke/clouds") burned cannabis flowers to induce trance.

Cannabis sativa from Vienna Dioscurides, 512 AD

Cannabis has an ancient history of ritual use and is found in pharmacological cults around the world. Hemp seeds discovered by archaeologists at Pazyryk suggest early ceremonial practices like eating by the Scythians occurred during the 5th to 2nd century BCE, confirming previous historical reports by Herodotus. It was used by Muslims in various Sufi orders as early as the Mamluk period, for example by the Qalandars. Smoking pipes uncovered in Ethiopia and carbon-dated to around 1320 CE were found to have traces of cannabis.

A study published in the *South African Journal of Science* showed that "pipes dug up from the garden of Shakespeare's home in Stratford-upon-Avon contain traces of cannabis." The chemical analysis was carried out after researchers hypothesized that the "noted weed" mentioned in Sonnet 76 and the "journey in my head" from Sonnet 27 could be references to cannabis and the use thereof. Examples of classic literature featuring cannabis include *Les paradis artificiels* by Charles Baudelaire and *The Hasheesh Eater* by Fitz Hugh Ludlow.

In the mid-1800s, Irish physician William Brooke O'Shaughnessy, who had studied the drug while working as a medical officer in Bengal with the East India company, brought a quantity of cannabis with him on his return to Britain in 1842, provoking renewed interest in the West.

John Gregory Bourke described the use of "mariguan", which he identifies as *Cannabis indica* or Indian hemp, by Mexican residents of the Rio Grande region of Texas in 1894. He described its uses for treatment of asthma, to expedite delivery, to keep away witches, and as a love-philtre. He also wrote that many Mexicans added the herb to their cigarritos or mescal, often taking a bite of sugar afterward to intensify the effect. Bourke wrote that because it was often used in a mixture with toloachi (which he inaccurately describes as *Datura stramonium*), mariguan was one of the several plants known as "loco weed". Bourke compared mariguan to hasheesh, which he called "one of the greatest curses of the East", citing reports that users "become maniacs and are apt to commit all sorts of acts of violence and murder", causing degeneration of the body and an idiotic appearance, and mentioned laws against sale of hasheesh "in most Eastern countries".

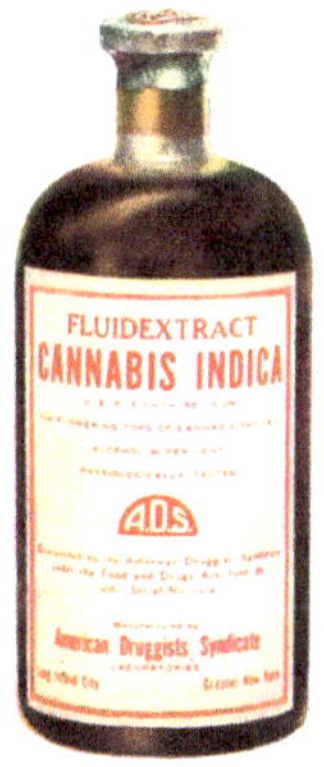

Cannabis indica fluid extract, American Druggists Syndicate, pre-1937

Cannabis was criminalized in various countries beginning in the early 20th century. In the United States, the first restrictions on sale of cannabis came in 1906 (in District of Columbia). It was outlawed in Jamaica (then a British colony) in 1913, in South Africa in 1922, and in the United Kingdom and New Zealand in the 1920s. Canada criminalized cannabis in the Opium and Drug Act of 1923, before any reports of the use of the drug in Canada. In 1925 a compromise was made at an international conference in The Hague about the International Opium Convention that banned exportation of "Indian hemp" to countries that had prohibited its use, and requiring importing countries to issue certificates approving the importation and stating that the shipment was required "exclusively for medical or scientific purposes". It also required parties to "exercise an effective control of such a nature as to prevent the illicit international traffic in Indian hemp and especially in the resin".

In the United States in 1937, the Marihuana Tax Act was passed, and prohibited the production of hemp in addition to cannabis. The reasons that hemp was also included in this law are disputed—several scholars have claimed that the act was passed in order to destroy the US hemp industry, with the primary involvement of businessmen Andrew Mellon, Randolph Hearst, and the Du Pont family. But the improvements of the decorticators, machines that separate the fibers from the hemp stem, could not make hemp fiber a very cheap substitute for fibers from other sources because it could not change that basic fact that strong fibers are only found in the bast, the outer part of the stem. Only about 1/3 of the stem are long and strong fibers. The company DuPont and many industrial historians dispute a link between nylon and hemp. They argue that the purpose of developing the nylon was to produce a fiber that could be used in thin stockings for females and compete with silk.

The United Nations' 2012 *Global Drug Report* stated that cannabis "was the world's most widely produced, trafficked, and consumed drug in the world in 2010", identifying that between 119 million and 224 million users existed in the world's adult (18 or older) population.

Society and Culture

Global estimates of illegal drug users in 2014 (in millions of users)			
Substance	**Best estimate**	**Low estimate**	**High estimate**
Amphetamine-type stimulants	35.65	15.34	55.90
Cannabis	182.50	127.54	233.65
Cocaine	18.26	14.88	22.08
Ecstasy	19.40	9.89	29.01
Opiates	17.44	13.74	21.59
Opioids	33.12	28.57	38.52

Legal Status

Cannabis propaganda sheet from 1935

Since the beginning of the 20th century, most countries have enacted laws against the cultivation, possession or transfer of cannabis. These laws have impacted adversely on the cannabis plant's cultivation for non-recreational purposes, but there are many regions where, under certain circumstances, handling of cannabis is legal or licensed. Many jurisdictions have lessened the penalties for possession of small quantities of cannabis so that it is punished by confiscation and sometimes a fine, rather than imprisonment, focusing more on those who traffic the drug on the black market.

In some areas where cannabis use has been historically tolerated, some new restrictions have been put in place, such as the closing of cannabis coffee shops near the borders of the Netherlands, closing of coffee shops near secondary schools in the Netherlands and crackdowns on "Pusher Street" in Christiania, Copenhagen in 2004.

Some jurisdictions use free voluntary treatment programs and/or mandatory treatment programs for frequent known users. Simple possession can carry long prison terms in some countries, particularly in East Asia, where the sale of cannabis may lead to a sentence of life in prison or even execution. More recently, however, many political parties, non-profit organizations, and causes based on the legalization of medical cannabis and/or legalizing the plant entirely (with some restrictions) have emerged.

In December 2012, the U.S. state of Washington became the first state to officially legalize cannabis in a state law (Washington Initiative 502) (but still illegal by federal law), with the state of Colorado following close behind (Colorado Amendment 64). On January 1, 2013, the first marijuana "club" for private marijuana smoking (no buying or selling, however) was allowed for the first time in Colorado. The California Supreme Court decided in May 2013 that local governments can ban medical marijuana dispensaries despite a state law in California that permits the use of cannabis for medical purposes. At least 180 cities across California have enacted bans in recent years.

In December 2013, Uruguay became the first country to legalize growing, sale and use of cannabis. However, as of August 2014, no cannabis has yet been sold legally in Uruguay. According to the law, the only cannabis that can be sold legally must be grown in the country by no more than five licensed growers, and these have yet to be selected; in fact, the call for applications did not go out until August 1, 2014.

On October 17, 2015, Australian health minister Sussan Ley presented a new law that will allow the cultivation of cannabis for scientific research and medical trials on patients. In December 2015, it was reported that the Canadian government had committed to legalizing cannabis, but at that time no timeline for the legalization was set out.

As drugs has increasingly come to be seen as a health issue instead of criminal behavior, marijuana has also been legalized or decriminalized in: Czech Republic, Colombia, Ecuador, Mexico, Portugal, and Canada.

Usage

In 2013, between 128 and 232 million people used cannabis (2.7% to 4.9% of the global population between the ages of 15 and 65). Cannabis is by far the most widely used illicit substance.

United States

Between 1973 and 1978, eleven states decriminalized marijuana. In 2001 Nevada reduced marijuana possession to a misdemeanor and since 2012, several other states have decriminalized and even legalized marijuana.

In 2015, almost half of the people in the United States had tried marijuana, 12% had used it in the past year, and 7.3% had used it in the past month. In 2014, daily marijuana use amongst US college students had reached its highest level since records began in 1980, rising from 3.5% in 2007 to 5.9% in 2014 and had surpassed daily cigarette use.

In the US, men are over twice as likely to use marijuana as women and 18-29 year-olds are six times more likely to use as over 65-year-olds. In 2015, a record 44% of the US population has tried marijuana in their lifetime, an increase from 38% in 2013 and 33% in 1985.

Marijuana use in the United States is three times above the global average, but in line with other Western democracies. 44% of American 12th graders have tried the drug at least once, and the typical age of first-use is 16, similar to the typical age of first-use for alcohol but lower than the first-use age for other illicit drugs.

Economics

Woman selling cannabis and bhang in Guwahati, Assam, India

Production

It is often claimed by growers and breeders of herbal cannabis that advances in breeding and cultivation techniques have increased the potency of cannabis since the late 1960s and early '70s when THC was first discovered and understood. However, potent seedless cannabis such as "Thai sticks" were already available at that time. *Sinsemilla* (Spanish for "without seed") is the dried, seedless inflorescences of female cannabis plants. Because THC production drops off once pollination occurs, the male plants (which produce little THC themselves) are eliminated before they shed pollen to prevent pollination. Advanced cultivation techniques such as hydroponics, cloning, high-intensity artificial lighting, and the sea of green method are frequently employed as a response (in part) to prohibition enforcement efforts that make outdoor cultivation more risky. It is often cited that the average levels of THC in cannabis sold in the United States rose dramatically between the 1970s and 2000, but such statements are likely skewed because undue weight is given to much more expensive and potent, but less prevalent samples.

"Skunk" refers to several named strains of potent cannabis, grown through selective breeding and sometimes hydroponics. It is a cross-breed of *Cannabis sativa* and *C. indica* (although other strains of this mix exist in abundance). Skunk cannabis potency ranges usually from 6% to 15% and rarely as high as 20%. The average THC level in coffee shops in the Netherlands is about 18–19%.

Price

The price or street value of cannabis varies widely depending on geographic area and potency.

In the United States, cannabis is overall the number four value crop, and is number one or two in many states including California, New York and Florida, averaging $3,000 per pound ($6,600/kg). Some believe it generates an estimated $36 billion market. Some have argued that this estimate is methodologically flawed, and makes unrealistic assumptions about the level of marijuana consumption. Other estimates claiming to correct for this flaw claim that the market is between $2.1-$4.3 billion. The United Nations Office on Drugs and Crime claims in its 2008 World Drug Report that typical U.S. retail prices are $10–15 per gram (approximately $280–420 per ounce). Street prices in North America are known to range from about $40–$400 per ounce ($1.4–$14/g), depending on quality.

The European Monitoring Centre for Drugs and Drug Addiction reports that typical retail prices in Europe for cannabis varies from €2 to €20 per gram, with a majority of European countries reporting prices in the range €4–10.

Gateway Drug

The Gateway Hypothesis states that cannabis use increases the probability of trying "harder" drugs. The hypothesis has been hotly debated as it is regarded by some as the primary rationale for the United States prohibition on cannabis use. A Pew Research Center poll found that political opposition to marijuana use was significantly associated with concerns about health effects and whether legalization would increase marijuana use by children.

Some studies state that while there is no proof for the gateway hypothesis, young cannabis users should still be considered as a risk group for intervention programs. Other findings indicate that hard drug users are likely to be poly-drug users, and that interventions must address the use of multiple drugs instead of a single hard drug. Almost two-thirds of the poly drug users in the "2009/10 Scottish Crime and Justice Survey" used cannabis.

The gateway effect may appear due to social factors involved in using any illegal drug. Because of the illegal status of cannabis, its consumers are likely to find themselves in situations allowing them to acquaint with individuals using or selling other illegal drugs. Utilizing this argument some studies have shown that alcohol and tobacco may additionally be regarded as gateway drugs; however, a more parsimonious explanation could be that cannabis is simply more readily available (and at an earlier age) than illegal hard drugs. In turn alcohol and tobacco are easier to obtain at an earlier point than is cannabis (though the reverse may be true in some areas), thus leading to the "gateway sequence" in those individuals since they are most likely to experiment with any drug offered.

An alternative to the gateway hypothesis is the common liability to addiction (CLA) theory. It states that some individuals are, for various reasons, willing to try multiple recreational substances. The "gateway" drugs are merely those that are (usually) available at an earlier age than the harder drugs. Researchers have noted in an extensive review, Vanyukov et al., that it is dangerous to present the sequence of events described in gateway "theory" in causative terms as this hinders both research and intervention.

Research

Cannabis research is challenging since the plant is illegal in most countries. Research-grade samples of the drug are difficult to obtain for research purposes, unless granted under authority of national governments.

There are also other difficulties in researching the effects of cannabis. Many people who smoke cannabis also smoke tobacco. This causes confounding factors, where questions arise as to whether the tobacco, the cannabis, or both that have caused a cancer. Another difficulty researchers have is in recruiting people who smoke cannabis into studies. Because cannabis is an illegal drug in many countries, people may be reluctant to take part in research, and if they do agree to take part, they may not say how much cannabis they actually smoke.

A 2015 review found that the use of high cannabidiol-to-THC strains of cannabis showed significantly fewer positive symptoms such as delusions and hallucinations, better cognitive function and both lower risk for developing psychosis, as well as a later age of onset of the illness, compared to cannabis with low CBD-to-THC ratios. A 2014 Cochrane review found that research was insufficient to determine the safety and efficacy to using cannabis to treat schizophrenia or psychosis.

Adaptogen

Panax quinquefolius or ginseng is often claimed to have adaptogenic effects.

Adaptogens or adaptogenic substances, compounds, herbs or practices refer to the pharmacological concept whereby administration results in stabilization of physiological processes and promotion of homeostasis, for example, decreased cellular sensitivity to stress. The European Medicines Agency states in a 2008 reflection paper that the concept requires additional clinical and pre-clinical research, and is therefore not accepted into current terminology.

Normal pharmacological theory supports the receptor theory of drug mechanism, and the scientific community calls into question the validity of the concept of adaptogens, while the U.S. Food and Drug Administration (FDA) and European Medicines Agency accept that the term can be used to assess traditional products.

In herbal medicine the categorization of different herbs as adaptogens is very popular, often with far-reaching claims of increasing longevity, libido and well-being. Most of the studies conducted on adaptogens were performed in the Soviet Union, Korea, and China during the 1980's. These studies have been criticized for various methodological flaws.

Adaptogens have been claimed to treat a wide variety of medical conditions, from fatigue to cancer.

History

The concept of adaptogens was originally created by pharmacologist N.V. Lazarev in 1947 to describe novel effects of dibazol 12-benzyl benzimidazol, an arterial dilator developed in France. Adaptogenesis was later applied in the former Soviet Union to describe remedies that increase the resistance of organisms to biological stress in experimental and clinical studies. According to the original definition, adaptogens are:

non-specific remedies *"that increase resistance to a broad spectrum of harmful factors (stressors) of different physical, chemical and biological natures"*.

This definition has been updated and today adaptogens are conceptualized as a:

"new class of metabolic regulators (of a natural origin) which increase the ability of an organism to adapt to environmental factors and to avoid damage from such factors."

Another definition requires that, for an herb to be considered an adaptogen, it must be nontoxic, nonspecific and have a normalizing effect on physiology.

By 1984, more than 1,500 pharmacological and clinical published studies had been conducted on adaptogens in the USSR, Adaptogens have even been described as "medicine[sic] for healthy people". In 1998, the term "adaptogen" was allowed as a functional claim for certain products by U.S. Food and Drug Administration and the European Medicines Agency claims, "The concept of adaptogens is sufficient to be considered in the assessment of traditional herbal medicinal products (e.g., monograph on Eleutherococcus root)." However they concluded saying "The HMPC is aware of the fact that numerous pre-clinical and clinical studies have been performed with the view to proving the concept of an adaptogen. However, the clinical data have a number of shortcomings such as deficiencies in the description of inclusion and exclusion criteria, description of the medication, diagnosis, study design, analysis etc.

Adaptogenic Plants

There is dispute in the herbal community regarding what qualifies as an adaptogen. However, the following are generally believed to qualify:

- American Ginseng (*Panax quinquefolius*), root
- Ashwagandha (*Withania somnifera*), root
- Asian Ginseng (*Panax ginseng*), root
- Cordyceps (*Cordyceps sinensis*), mushroom/mycelium
- Dang Shen (*Codonopsis pilosula, C. tangshen*), root
- Eleuthero (*Eleutherococcus senticosus*), root/stem bark
- Green Chirayta (*Andrographis paniculata*), leaves
- Guduchi (*Tinospora cordifolia*), root/stem
- Holy Basil (*Ocimum sanctum, O. gratissimum*), herb
- Jiaogulan (*Gynostemma pentaphyllum*), herb
- Licorice (*Glycyrrhiza glabra, G. uralensis*), root
- Reishi (*Ganoderma ludicum*), mushroom/mycelium
- Rhaponticum (*Rhaponticum carthamoides*), root
- Rhodiola (*Rhodiola rosea*), root
- Schisandra (*Schisandra chinensis*), fruit/seed

- Siberian Ginseng (*Eleutherococcus Senticosus*), root,
- Shilajit (*Asphaltum bitumen*), pitch
- Shiitake (*Lentinula edodes*)

Adaptogens in Other Cultures

The concept of adaptogens as "medicine for the healthy" or as a substance that helps the body cope with stress is a similar to many remedies common in chinese herbology, as well as other forms of traditional medicine.

Ayurveda traditional medicine operates with the concept of rasayana. Various substances are classified in this tradition as rasayanas, meaning they are believed to promote physical and mental health, improve defense mechanisms of the body and enhance longevity. Rasayanas are referred to as adaptogens by some researchers.

Aromatherapy

Aromatherapy uses plant materials and aromatic plant oils, including essential oils, and other aroma compounds for improving psychological or physical well-being.

It can be offered as a complementary therapy or, more controversially, as a form of alternative medicine. Complementary therapy can be offered alongside standard treatment, with alternative medicine offered instead of conventional, evidence-based treatments.

Aromatherapists, who specialize in the practice of aromatherapy, utilize blends of therapeutic essential oils that can be issued through topical application, massage, inhalation or water immersion to stimulate a desired response.

There is no good medical evidence that aromatherapy can either prevent or cure any disease, but it might help improve general well-being.

History

The use of essential oils for therapeutic, spiritual, hygienic and ritualistic purposes goes back to a number of ancient civilizations including the Chinese, Indians, Egyptians, Greeks, and Romans who used them in cosmetics, perfumes and drugs.

Oils are described by Dioscorides, along with beliefs of the time regarding their healing properties, in his *De Materia Medica*, written in the first century. Distilled essential oils have been employed as medicines since the invention of distillation in the eleventh century, when Avicenna isolated essential oils using steam distillation.

The concept of aromatherapy was first mooted by a small number of European scientists and doctors, in about 1907. In 1937, the word first appeared in print in a French book on the subject: *Aromathérapie: Les Huiles Essentielles, Hormones Végétales* by René-Maurice Gattefossé, a chemist. An English version was published in 1993. In 1910, Gattefossé burned a hand very badly and later claimed he treated it effectively with lavender oil.

A French surgeon, Jean Valnet, pioneered the medicinal uses of essential oils, which he used as antiseptics in the treatment of wounded soldiers during World War II.

Modes of Application

The modes of application of aromatherapy include:

- *Aerial diffusion*: for environmental fragrancing or aerial disinfection
- *Direct inhalation*: for respiratory disinfection, decongestant, expectoration as well as psychological effects
- *Topical applications*: for general massage, baths, compresses, therapeutic skin care

Materials

Some of the materials employed include:

- Absolutes: fragrant oils extracted primarily from flowers or delicate plant tissues through solvent or supercritical fluid extraction (e.g., rose absolute). The term is also used to describe oils extracted from fragrant butters, concretes, and enfleurage pommades using ethanol.
- Aroma lamps or diffusers: an electric or candle-fueled device which volatilizes essential oils, usually mixed with water.
- Carrier oils: typically oily plant base triacylglycerides that dilute essential oils for use on the skin (e.g., sweet almond oil).
- Essential oils: fragrant oils extracted from plants chiefly through steam distillation (e.g., eucalyptus oil) or expression (grapefruit oil). However, the term is also occasionally used to describe fragrant oils extracted from plant material by any solvent extraction. This material includes incense reed diffusers.
- Herbal distillates or hydrosols: the aqueous by-products of the distillation process (e.g., rosewater). Common herbal distillates are chamomile, rose, and lemon balm.
- Infusions: aqueous extracts of various plant material (e.g., infusion of chamomile).
- Phytoncides: various volatile organic compounds from plants that kill microbes. Many terpene-based fragrant oils and sulfuric compounds from plants in the genus "*Allium*" are phytoncides, though the latter are likely less commonly used in aromatherapy due to their disagreeable odors.
- Vaporizers: typically higher oil content plant based materials dried, crushed, and heated to extract and inhale the aromatic oil vapors in a direct inhalation modality.

Theory

Aromatherapy is the treatment or prevention of disease by use of essential oils. Other stated uses include pain and anxiety reduction, enhancement of energy and short-term memory, relaxation, hair loss prevention, and reduction of eczema-induced itching.

Two basic mechanisms are offered to explain the purported effects. One is the influence of aroma on the brain, especially the limbic system through the olfactory system. The other is the direct pharmacological effects of the essential oils.

In the English-speaking world, practitioners tend to emphasize the use of oils in massage. Aromatherapy tends to be regarded as a pseudoscientific fraud at worst.

Choice and Purchase

Oils with standardized content of components (marked FCC, for Food Chemicals Codex) are required to contain a specified amount of certain aroma chemicals that normally occur in the oil. There is no law that the chemicals cannot be added in synthetic form to meet the criteria established by the FCC for that oil. For instance, lemongrass essential oil must contain 75% aldehyde to meet the FCC profile for that oil, but that aldehyde can come from a chemical refinery instead of from lemongrass. To say that FCC oils are "food grade" makes them seem natural when they are not necessarily so.

Undiluted essential oils suitable for aromatherapy are termed 'therapeutic grade', but there are no established and agreed standards for this category.

Analysis using gas liquid chromatography (GLC) and mass spectrometry (MS) establishes the quality of essential oils. These techniques are able to measure the levels of components to a few parts per billion. This does not make it possible to determine whether each component is natural or whether a poor oil has been 'improved' by the addition of synthetic aromachemicals, but the latter is often signaled by the minor impurities present. For example, linalool made in plants will be accompanied by a small amount of hydro-linalool, whilst synthetic linalool has traces of dihydro-linalool.

Effectiveness

There is some evidence that aromatherapy can help improve general well-being, but no good medical evidence that it can prevent or cure disease.

In 2015 the Australian Government's Department of Health published the results of a review of alternative therapies that sought to determine if any were suitable for being covered by health insurance; Aromatherapy was one of 17 therapies evaluated for which no clear evidence of effectiveness was found. Evidence for the efficacy of aromatherapy in treating medical conditions is poor, with a particular lack of studies employing rigorous methodology.

Safety Concerns

Aromatherapy carries a risk of a number of adverse effects and this consideration, combined with the lack of evidence of its therapeutic benefit, makes the practice of questionable worth.

Because essential oils are highly concentrated they can irritate the skin when used in undiluted form. Therefore, they are normally diluted with a carrier oil for topical application, such as jojoba oil, olive oil, or coconut oil. Phototoxic reactions may occur with citrus peel oils such as lemon or lime. Also, many essential oils have chemical components that are sensitisers (meaning that they will, after a number of uses, cause reactions on the skin, and more so in the rest of the body). Some of the chemical allergies could even be caused by pesticides, if the original plants are cultivated. Some oils can be toxic to some domestic animals, with cats being particularly prone.

A child hormone specialist at the University of Cambridge claimed "... these oils can mimic estrogens" and "people should be a little bit careful about using these products." The Aromatherapy Trade Council of the UK has issued a rebuttal. The Australian Tea Tree Association, a group that promotes the interests of Australian tea tree oil producers, exporters and manufacturers issued a letter that questioned the study and called on the *New England Journal of Medicine* for a retraction. The *New England Journal of Medicine* has so far not replied and has not retracted the study.

As with any bioactive substance, an essential oil that may be safe for the general public could still pose hazards for pregnant and lactating women.

While some advocate the ingestion of essential oils for therapeutic purposes, licensed aromatherapy professionals do not recommend self-prescription due to the highly toxic nature of some essential oils. Some very common oils like eucalyptus are extremely toxic when taken internally. Doses as low as one teaspoon have been reported to cause clinically significant symptoms and severe poisoning can occur after ingestion of 4 to 5 ml. A few reported cases of toxic reactions like liver damage and seizures have occurred after ingestion of sage, hyssop, thuja, and cedar. Accidental ingestion may happen when oils are not kept out of reach of children.

Oils both ingested and applied to the skin can potentially have negative interactions with conventional medicine. For example, the topical use of methyl salicylate-heavy oils like sweet birch and wintergreen may cause bleeding in users taking the anticoagulant warfarin.

Adulterated oils may also pose problems depending on the type of substance used.

Cannabis Sativa

Cannabis sativa is an annual herbaceous plant in the Cannabis genus. It is a member of a small, but diverse family of flowering plants of the Cannabaceae family. It has been cultivated throughout recorded history, used as a source of industrial fibre, seed oil, food, recreation, religious and spiritual moods and medicine. Each part of the plant is harvested differently, depending on the purpose of its use. The species was first classified by Carl Linnaeus in 1753.

Plant Physiology

The flowers of *Cannabis sativa* are unisexual and plants are most often either male or female. It is a short-day flowering plant, with staminate (male) plants usually taller and less robust than pistillate (female) plants The flowers of the female plant are arranged in racemes and can produce hundreds of seeds. Male plants shed their pollen and die several weeks prior to seed ripening on the female plants. Although genetic factors dispose a plant to become male or female, environmental factors including the diurnal light cycle can alter sexual expression. Naturally occurring monoecious plants, with both male and female parts, are either sterile or fertile but artificially induced "hermaphrodites" can have fully functional reproductive organs. "Feminized" seed sold by many commercial seed suppliers are derived from artificially "hermaphrodytic" females that lack the male gene, or by treating the plants with hormones or silver thiosulfate.

Cultivars

Broadly, there are three main cultivar groups of cannabis that are cultivated today:

- Cultivars primarily cultivated for their fibre, characterised by long stems and little branching.
- Cultivars grown for seed which can be eaten entirely raw or from which hemp oil is extracted.
- Cultivars grown for medicinal or recreational purposes. A nominal if not legal distinction is often made between industrial hemp, with concentrations of psychoactive compounds far too low to be useful for that purpose, and marijuana.

Pharmacology

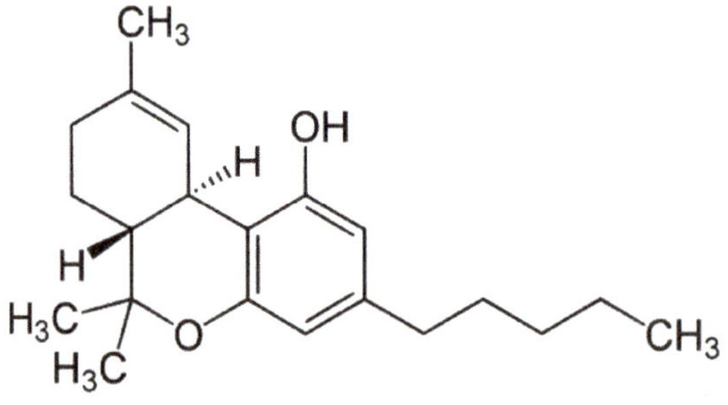

Δ^9-tetrahydrocannabinol (THC)

The flower of a hybrid Cannabis indica plant

Cannabis sativa, scientific drawing from c1900

Although the main psychoactive constituent of *Cannabis* is tetrahydrocannabinol (THC), the plant is known to contain more than 500 compounds, among them at least 113 cannabinoids; however, most of these "minor" cannabinoids are only produced in trace amounts. Besides THC, another cannabinoid produced in high concentrations by some plants is cannabidiol (CBD), which is not psychoactive but has recently been shown to block the effect of THC in the nervous system. Differences in the chemical composition of *Cannabis* varieties may produce different effects in humans. Synthetic THC, called dronabinol, does not contain CBD, CBN, or other cannabinoids, which is one reason why its pharmacological effects may differ significantly from those of natural *Cannabis* preparations.

Chemical Constituents

Beside cannabinoids, Cannabis chemical constituents include about 120 compounds responsible for its characteristic aroma. These are mainly volatile terpenes and sesquiterpenes.

- α-Pinene
- Myrcene
- Linalool
- Limonene
- Trans-β-ocimene
- α-Terpinolene
- Trans-caryophyllene
- α-Humulene, contributes to the characteristic aroma of *Cannabis sativa*
- Caryophyllene, with which some hashish detection dogs are trained

Difference Between C. Sativa and C. Indica

Some studies indicate that, on average, *Cannabis sativa* has higher levels of CBD compared to THC than *C. indica*. However, large variability exists within either species, and there is an increasing discussion whether the existing paradigm of the difference between species adequately represents the variability found within the *Cannabis* genus. There are five chemotaxonomic types of *Cannabis*: one with high levels of THC, one which is more fibrous and has higher levels of CBD, one that is an intermediate between the two, another one with high levels of cannabigerol (CBG), and the last one almost without cannabinoids.

Cannabis strains with relatively high CBD:THC ratios are less likely to induce anxiety than vice versa. This may be due to CBD's antagonistic effects at the cannabinoid receptors, compared to THC's partial agonist effect. CBD is also a 5-HT_{1A} receptor (serotonin) agonist, which may also contribute to an anxiolytic-content effect. The effects of *sativa* are well known for its cerebral high, while *indica* is well known for its sedative effects which some prefer for night time use. Both types are used as medical cannabis. *Indica* plants are normally shorter and stockier than *sativas*. They have broader, deeply serrated leaves and a compact and dense flower cluster.

Common uses

A sack made from hemp fibre

Cannabis sativa seeds are chiefly used to make hempseed oil which can be used for cooking, lamps, lacquers or paints. They can also be used as caged-bird feed, as they provide a moderate source of nutrients for most birds. The flowers (and to a lesser extent the leaves, stems, and seeds) contain psychoactive chemical compounds known as cannabinoids that are consumed for recreational, medicinal and spiritual purposes. When so used, preparations of flowers (marijuana) and leaves and preparations derived from resinous extract (e.g., hashish) are consumed by smoking, vaporising and oral ingestion. Historically, tinctures, teas and ointments have also been common preparations. In traditional medicine of India in particular C. sativa has been used as hallucinogenic, hypnotic, sedative, analgesic and anti-inflammatory agent.

Cultivation

In the case of production related to recreational use of Cannabis, the process of "cloning" provides the grower with DNA identical female plants eliminating the need of seeds for the growing process.

A *Cannabis* plant in the vegetative growth phase of its life requires more than 12–13 hours of light per day to stay vegetative. Flowering usually occurs when darkness equals at least 12 hours per day. The flowering cycle can last anywhere between nine and fifteen weeks, depending on the strain and environmental conditions.

In soil, the optimum pH for the plant is 6.3 to 6.8. In hydroponic growing, the nutrient solution is best at 5.2 to 5.8, making *Cannabis* well-suited to hydroponics because this pH range is hostile to most bacteria and fungi.

Menerba

Menerba, also known as Menopause Formula 101 (MF-101), is a botanical drug candidate that acts as a selective estrogen receptor modulator (SERM) which is being studied for its potential to relieve hot flashes associated with menopause. Menerba, an estrogen receptor beta (ERβ) agonist (ERBA), is part of a new class of receptor subtype-selective estrogens, which is selective in transcriptional regulation to one of the two known estrogen receptor (ER) subtypes. Menerba consists of 22 herbs that have been used historically in traditional Chinese medicine.

Menerba binds to both ERα and ERβ and with equal affinity, but does not activate ERα and instead activates only ERβ-mediated gene transcription.

Mode of Action

Menerba has been reported to alleviate menopausal symptoms such as hot flashes, while having no stimulative effects on endometrium or breast tissue.

In mouse xenograft models, Menerba produced a different conformation in estrogen receptor alpha (ERα) from ERβ when compared with the conformations produced by estradiol. The specific conformational change induced by Menerba allows ERβ to bind to an estrogen response element and recruit coregulatory proteins that are required for gene activation. It has been shown that the increased risk of breast and uterine cancers is associated with ERα activation and that ERβ blocks the growth promoting effects on breast cancer cells. Menerba does not activate the ERα-regulated proliferative genes, c-myc and cyclin D1, or stimulate MCF-7 breast cancer cell proliferation or tumor formation, demonstrating that it may be a viable alternative for hormone therapy in comparison to estrogens that non-selectively activate both ER subtypes.

In 2007, Menerba completed a multi-center Phase 2, double-blind, placebo-controlled randomized clinical trial evaluating its potential for the treatment of hot flashes in 217 healthy post-menopausal women in the U.S. The principal investigator of the trial was Dr. Deborah Grady from the University of California, San Francisco. Menerba showed a statistically significant reduction in the number of hot flashes after 12 weeks of treatment and had a statistically significant reduction in nighttime awakenings from hot flashes (-67%, p=0.05). There was no difference in uterine bleeding between treatment groups and placebo, and no uterine abnormalities were observed during the study. The only side effect observed was mild loose stools (12% in treatment group vs. 3% in the placebo group).

Further studies have shown that most active estrogenic compound in Menerba is liquiritigenin, derived from the root of *Glycyrrhizae uralensis* Fisch, one of the 22 botanically derived ingredients found in Menerba. In a mouse xenograft model, liquiritigenin activated multiple ER regulatory elements and native target genes with ERβ but not ERα. The ERβ-selectivity of liquiritigenin was due to the selective recruitment of the coactivator steroid receptor coactivator-2 to target genes. Liquiritigenin did not stimulate uterine size or tumorigenesis of MCF-7 breast cancer cells. The results demonstrate that some plants contain highly selective estrogens for ERβ and are as selective as synthetic compounds, but regulate different genes. (PLOs 1 – July 17) and suggests that plant-derived ERb-selective compounds could lead to safer, more attractive alternative therapies for menopausal symptoms. In a further study, Menerba was shown to regulate calcium influx, which is related to temperature regulation.

Menerba underwent Phase 1 safety trials evaluating its toxicity at 10 g/day and 15 g/day doses. No significant toxicities have been observed in any of the animal studies with doses ranging from 2000 mg/kg/day in dogs to 16,000 mg/kg/day in rodents. Due to acceptable toxicity results on the Phase 1 trial, the FDA approved Menerba to initiate a Phase 3 trial with 1200 women reporting seven or more menopausal hot flashes per day.

Decoction

Turkish coffee beginning to boil. Decoction compares to brewing coffee through percolation.

Decoction is a method of extraction by boiling herbal or plant material to dissolve the chemicals of the material, which may include stems, roots, bark and rhizomes. Decoction involves first mashing the plant material to allow for maximum dissolution, and then boiling in water to extract oils, volatile organic compounds and other various chemical substances. Decoction can be used to make herbal teas, leaf teas, coffees, tinctures and similar solutions. Decoctions and infusions may produce liquids with differing chemical properties as the temperature and/ or preparation difference may result in more oil-soluble chemicals in decoctions versus infusions. The process can also be applied to meats and vegetables to prepare bouillon or stock, though the term is typically only used to describe boiled plant extracts, usually for medicinal or scientific purposes.

A decoction is also the name for the resulting liquid. Although this method of extraction differs from infusion and percolation, the resultant liquids can sometimes be similar in their effects, or general appearance and taste.

Etymology

The term dates back to 1350–1400 from present participle stem of Latin *decoquere* (meaning to boil down), de “from” + coquere “to cook”.

Use

In brewing, decoction mashing is the traditional method where a portion of the mash is removed to a separate vessel, boiled for a time and then returned to the main mash, raising the mash to the next temperature step.

In herbalism, decoctions are usually made to extract fluids from hard plant materials such as roots and bark. To achieve this, the plant material is usually boiled for 1–2 hours in 1-5 liters of water. It is then strained.

References

- Capasso, F., Gaginella, T.S., Grandolini, G., Izzo, A.A. (2003). Phytotherapy A Quick Reference to Herbal Medicine. Springer. ISBN 978-3-540-00052-5
- Hong, Francis (2004). “History of Medicine in China” (PDF). McGill Journal of Medicine. 8 (1): 7984. Archived from the original (PDF) on 2013-12-01
- “Herbs at a Glance”. National Center for Complementary and Integrative Health, US National Institutes of Health. 21 November 2016. Retrieved 24 February 2017
- Cravotto G, Boffa L, Genzini L, Garella D (February 2010). “Phytotherapeutics: an evaluation of the potential of 1000 plants”. J Clin Pharm Ther. 35 (1): 11–48. PMID 20175810. doi:10.1111/j.1365-2710.2009.01096.x
- Robson, Barry & Baek, O.K. (2009). The Engines of Hippocrates: From the Dawn of Medicine to Medical and Pharmaceutical Informatics. John Wiley & Sons. p. 50. ISBN 9780470289532
- Edgar J. DaSilva; Elias Baydoun; Adnan Badran (2002). “Biotechnology and the developing world”. Electronic Journal of Biotechnology. 5 (1). ISSN 0717-3458. doi:10.2225/vol5-issue1-fulltext-1
- Carroll, Lucy (24 February 2014). “Herbal medicines: Study raises alarm over labelling”. The Sydney Morning Herald, Australia. Retrieved 25 February 2017
- Burstein, H. J.; Gelber, S; Guadagnoli, E; Weeks, J. C. (June 1999). “Use of alternative medicine by women with early-stage breast cancer”. N Engl J Med. 340 (22): 1733–9. PMID 10352166. doi:10.1056/NEJM199906033402206
- Unschuld, Pual (2003). Huang Di Nei Jing: Nature, Knowledge, Imagery in an Ancient Chinese Medical Text. University of California Press. p. 286. ISBN 978-0-520-92849-7

- Gaire, B. P.; Subedi, L (2013). "A review on the pharmacological and toxicological aspects of Datura stramonium L". Journal of Integrative Medicine. 11 (2): 73–9. PMID 23506688. doi:10.3736/jintegrmed2013016
- "Botanical Dietary Supplements". Office of Dietary Supplements, US National Institutes of Health. June 2011. Retrieved 25 February 2017
- Müller, JL (1998). "Love potions and the ointment of witches: historical aspects of the nightshade alkaloids". J Toxicol Clin Toxicol. 36 (6): 617–27. PMID 9776969. doi:10.3109/15563659809028060
- Green, James (2000). The Herbal Medicine Maker's Handbook: A Home Manual. Chelsea Green Publishing. p. 168. ISBN 9780895949905
- Ernst E (1998). "Harmless Herbs? A Review of the Recent Literature" (PDF). The American Journal of Medicine. 104 (2): 170–8. PMID 9528737. doi:10.1016/S0002-9343(97)00397-5. Retrieved 27 December 2010
- Pettitt, Paul B. (2002). "The Neanderthal Dead, exploring mortuary variability in middle paleolithic eurasia." (PDF). Before Farming. 1 (4): 19 note 8. Retrieved December 29, 2015
- Williamson, E (2001). "Synergy and other interactions in phytomedicines". Phytomedicine. 8 (5): 401–409. PMID 11695885. doi:10.1078/0944-7113-00060
- Capasso, F., Gaginella, T.S., Grandolini, G., Izzo, A.A. (2003). Phytotherapy A Quick Reference to Herbal Medicine. Springer. ISBN 978-3-540-00052-5
- Cindy Jones. "Herbal Waters or Distillates (Hydrosols)". Sagescript Institute. Archived from the original on 2006-10-28. Retrieved 2006-10-23
- Breakspear I (2006). "Education and Regulation in Herbal Medicine: An Australian Perspective". Journal of the American Herbalists Guild. 6 (2): 35–38
- Sanie Lopez Garelli (25 November 2008). "Mexico, Paraguay top pot producers, U.N. report says". CNN International. Retrieved 28 September 2013
- Engel, Cindy (2002). Wild Health: How Animals Keep Themselves Well and What We Can Learn From Them. Houghton Mifflin. ISBN 0-618-07178-4
- Borges, G; Bagge, CL; Orozco, R (9 February 2016). "A literature review and meta-analyses of cannabis use and suicidality.". Journal of Affective Disorders. 195: 63–74. PMID 26872332. doi:10.1016/j.jad.2016.02.007
- "Brazilian Archives of Biology and Technology – Jurema-Preta (Mimosa tenuiflora [Willd.] Poir.): a review of its traditional use, phytochemistry and pharmacology". scielo.br. Retrieved 2009-01-14
- Dybas, Raskin (2007). "Out of Africa: A Tale of Gorillas, Heart Disease... and a Swamp Plant". BioScience. 57 (5): 392–397. doi:10.1641/B570503
- Susan G. Wynn; Barbara Fougère (2007). Veterinary Herbal Medicine. Elsevier Health Sciences. p. 60. ISBN 0323029981
- Tashkin, Donald (March 1997). "Effects of marijuana on the lung and its immune defenses". UCLA School of Medicine. Retrieved 2012-06-23
- Newton, David E. (2013). Marijuana : a reference handbook. Santa Barbara, Calif.: ABC-CLIO. p. 7. ISBN 9781610691499
- Truitt Elly R (2009). "The Virtues Of Balm In Late Medieval Literature". Early Science & Medicine. 14 (6): 711–736. doi:10.1163/138374209x12542104913966
- Allan Tasman; Jerald Kay; Jeffrey A. Lieberman; Michael B. First; Mario Maj (2011). Psychiatry. John Wiley & Sons. p. 9. ISBN 978-1-119-96540-4

Permissions

All chapters in this book are published with permission under the Creative Commons Attribution Share Alike License or equivalent. Every chapter published in this book has been scrutinized by our experts. Their significance has been extensively debated. The topics covered herein carry significant information for a comprehensive understanding. They may even be implemented as practical applications or may be referred to as a beginning point for further studies.

We would like to thank the editorial team for lending their expertise to make the book truly unique. They have played a crucial role in the development of this book. Without their invaluable contributions this book wouldn't have been possible. They have made vital efforts to compile up to date information on the varied aspects of this subject to make this book a valuable addition to the collection of many professionals and students.

This book was conceptualized with the vision of imparting up-to-date and integrated information in this field. To ensure the same, a matchless editorial board was set up. Every individual on the board went through rigorous rounds of assessment to prove their worth. After which they invested a large part of their time researching and compiling the most relevant data for our readers.

The editorial board has been involved in producing this book since its inception. They have spent rigorous hours researching and exploring the diverse topics which have resulted in the successful publishing of this book. They have passed on their knowledge of decades through this book. To expedite this challenging task, the publisher supported the team at every step. A small team of assistant editors was also appointed to further simplify the editing procedure and attain best results for the readers.

Apart from the editorial board, the designing team has also invested a significant amount of their time in understanding the subject and creating the most relevant covers. They scrutinized every image to scout for the most suitable representation of the subject and create an appropriate cover for the book.

The publishing team has been an ardent support to the editorial, designing and production team. Their endless efforts to recruit the best for this project, has resulted in the accomplishment of this book. They are a veteran in the field of academics and their pool of knowledge is as vast as their experience in printing. Their expertise and guidance has proved useful at every step. Their uncompromising quality standards have made this book an exceptional effort. Their encouragement from time to time has been an inspiration for everyone.

The publisher and the editorial board hope that this book will prove to be a valuable piece of knowledge for students, practitioners and scholars across the globe.

Index